T0227742

Rural Development in the Third World

The rural landscape of the Third World is generally seen as one worked by the impoverished. Chris Dixon shows that this is an increasingly inaccurate picture. Wealth does exist, with the landed often maintaining lifestyles comparable to their richest urban neighbours. And while land remains the basis of real wealth, the rural workforce is diversifying its activities away from agriculture becoming involved in a range of manufacturing, processing, trading and service industries. Yet still rural poverty persists, and the book illustrates just how difficult it is to assess the success of development initiatives adopted to eliminate it.

This book, first published in 1990, provides a general introduction to the approaches, policies, and problems associated with Third World rural development. *Rural Development in the Third World* will be of interest to students of geography, the environment and developmental issues.

Rural Development in the Third World

Chris Dixon

Routledge
Taylor & Francis Group

First published in 1990
by Routledge

This edition first published in 2015 by Routledge
2 Park Square, Milton Park, Abingdon, Oxon, OX14 4RN
and by Routledge
711 Third Avenue, New York, NY 10017

Routledge is an imprint of the Taylor & Francis Group, an informa business

© 1990 Chris Dixon

Publisher's Note
The publisher has gone to great lengths to ensure the quality of this reprint but points out that some imperfections in the original copies may be apparent.

Disclaimer
The publisher has made every effort to trace copyright holders and welcomes correspondence from those they have been unable to contact.

A Library of Congress record exists under LC control number: 89010390

ISBN 13: 978-1-138-92027-9 (hbk)
ISBN 13: 978-1-315-68575-5 (ebk)

Chris Dixon

Rural Development
in the Third World

R

ROUTLEDGE

London and New York

First published 1990
by Routledge
11 New Fetter Lane, London EC4P 4EE

Simultaneously published in the USA and Canada
by Routledge
a division of Routledge, Chapman and Hall, Inc.
29 West 35th Street, New York, NY 10001

Typeset by Witwell Ltd, Southport
Printed and bound in Great Britain by
Biddles Ltd, Guildford and King's Lynn

British Library Cataloguing in Publication Data

Dixon, C. J. (Christopher John, *1944-*)
    Rural development in the Third World. - (Routledge
    introductions to development)
    1. Developing countries. Rural regions. Economic
    development
    I. Title
    330.9172'4

ISBN 0-415-01597-9

Library of Congress Cataloguing in Publication Data

Dixon, C. J. (Chris J.)
    Rural development in the Third Word/Chris Dixon.
        p.   cm. — (Routledge introductions to development)
    Bibliography: p.
    Includex index.
    ISBN 0-415-01597-9
    1. Agriculture—Economic aspects—Developing countries. 2. Rural
development—Developing countries. I. Title.   II. Series.
HD1417.D57   1990
338.1'09172'4—dc20

Contents

Acknowledgements

The author wishes to thank Sue Shewan for typing the manuscript and the staff of the City of London Polytechnic Cartography Unit for the production of the maps, diagrams and photographs.

1
The rural sector

The rural areas of the Third World are usually described as the home of impoverished people engaged in agriculture. Whilst it is true that most rural people are poor, many are not even indirectly engaged in agriculture. Moreover, there are often some comparatively wealthy landowners, as well as people engaged in a range of manufacturing, processing, trading, and service activities. These are part of the rural sector and often have a significance out of proportion to their numbers. Some members of rural communities may even have income levels and life-styles comparable with those of the wealthiest urban dwellers.

While the main focus of this book is agriculture, the broader structure of the rural sector should not be forgotten. Land remains the main basis of wealth and political power in rural areas, but to classify rural people in terms of their access to land or involvement with agriculture can obscure the degree to which households engage in other activities and derive income from a wide and variable range of sources (plates 1.1 and 1.2). Figure 1.1 provides an illustration of the variations in household income sources during a year. For many rural households there will also be considerable variation from one year to the next.

Some writers refer to 'household survival strategies' which vary as people attempt to cope with changing conditions. Harvests, household food consumption, and market conditions are far from uniform. The amount of labour available will change because of births, deaths,

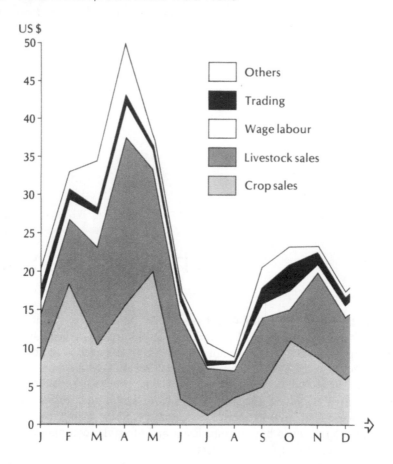

US $

Others

Trading

Wage labour

Livestock sales

Crop sales

J F M A M J J A S O N D

Figure 1.1 Monthly variations and composition of rural household cash income

marriages, illness, pregnancy, and migration. Cash needs will vary, as will the opportunities for earning. Thus 'snapshot' pictures of rural households based, for example, on single years (table 1.1) or individual months (figure 1.1) can produce very partial if not distorted impressions, since the period studied may be far from typical.

Large areas of the Third World are subject to a high degree of climatic seasonality, with distinct wet and dry seasons. This includes most of the Indian sub-continent, much of sub-Saharan Africa and extensive parts of South East Asia and Latin America. In the absence of irrigation the agricultural year and, indeed, most aspects of rural life are closely related

Plate 1.1 Stripping rattan for furniture manufacturing. A wide variety of rural industries provides permanent as well as part-time, temporary, and seasonal employment.

Table 1.1 Five years of rice cultivation in Thailand

	1978	1979	1980	1981	1982
Area planted (ha)	3·0	2·2	2·5	3·0	3·0
Area harvested (ha)	2·9	2·0	2·5	2·6	2·8
Production (kg)	5,400	2,600	4,625	3,640	4,480
Yield (kg/ha)	1,800	1,182	1,850	1,213	1,493
Adults	4	4	3	4	4
Children	3	4	3	4	3
Household rice need (kg)	2,310	2,310	1,980	2,310	2,310
Surplus rice for sale (kg)	3,090	290	2,645	1,330	2,170
Rice price (US$/10 kg)	0·25	0·35	0·30	0·25	0·20
Total household income (US$)	77·25	10·15	79·35	33·25	42·70
Income *per capita* (US$)	11·00	1·27	13·23	4·17	6·10
Rice production *per capita*	771·40	325·00	770·80	455·00	640·00
Planted land *per capita*	0·43	0·28	0·42	0·38	0·43

Plate 1.2 Spinning silk yarn. A variety of rural crafts produce goods for household consumption, exchange, or sale. They are often an important element in household survival strategies

Plate 1.3 In areas of seasonal rainfall large-scale storage of water is often
necessary

to the seasons (plate 1.3). The timing of a survey can, as is shown in case
study A, greatly affect the conclusions drawn from it. A high proportion of
studies are conducted in the period just after the harvest, since at this time
of year farmers are likely to be available to answer questions, the climate is
usually at its most pleasant, and access to communities easiest. Moreover,
as this is often the period of the year when conditions are most favourable,
rural poverty may be under-recorded.

The rural Third World in global perspective

As figure 1.2 indicates, in the Third World the proportion of the total
population living in rural areas is usually much greater than in the more
developed countries. There is in general an inverse relationship between
the proportion of people living in rural areas and *per capita* gross domestic
product (GDP) (table 1.2). A more detailed investigation, however reveals
a number of anomalies.

Apart from the city states of Hong Kong and Singapore, some Latin
American countries stand out as having a low proportion of the

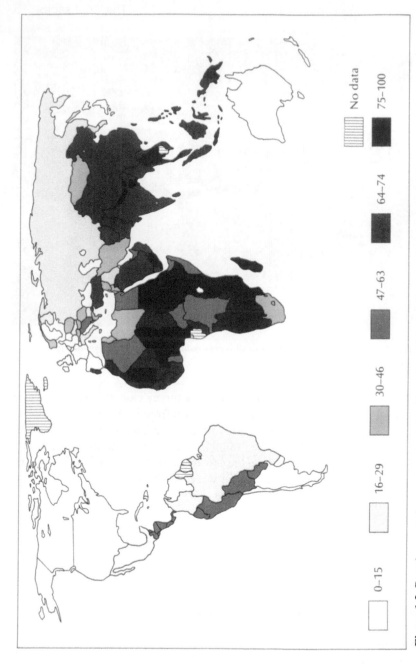

Figure 1.2 Rural populations as a proportion of total population. *Source:* World Bank (1988)

No data

75–100

64–74

47–63

30–46

16–29

0–15

population in rural areas for their level of development. In contrast, in Africa many sub-Saharan countries have some 90 per cent of their population living in rural areas. The variations between Third World countries are such that it is dangerous to read too much into the very general relationship illustrated in table 1.2.

Over the last twenty-five years there has been, in almost all Third World countries, a very substantial reduction in the proportion of people living in rural areas (see table 1.2). Generally the countries with the lowest *per capita* incomes have experienced the smallest changes. However, detailed investigation shows that some of the world's very poorest countries have experienced sharp falls in the rural sector's share of population. The causes of this are far from simple, though it is generally agreed that adverse conditions in rural areas are of prime importance in the process.

The only major exceptions to the general trend of declining rural share of the population are Vietnam and Kampuchea. During the Vietnam war these countries experienced a sharp increase in urbanization because of disruption in the countryside and, in the case of South Vietnam, forced migration to foster control and 'security'. The process was reversed from the mid-1970s, primarily owing to a deliberate policy of de-urbanization and rural development. As a result the rural sector's share of the population increased.

For most of the Third World while the proportion of the population living in rural areas has declined, the number has continued to increase. The absolute decline of rural populations evident in the United Kingdom from the late nineteenth century has only recently made a limited, and generally localized, appearance. Since the mid-1970s there have been sharp

Table 1.2 Percentage of the population living in rural areas and the level of *per capita* GDP, 1965–85

	1965	1985	Change
Low income countries	83	78	−5
Lower middle income countries	73	64	−7
Upper middle income countries	51	35	−16
All Third World countries	76	69	7

Note: In this and subsequent tables, low income = US$400 *per capita* or less; lower middle income = US$401–US$1,600 *per capita*; upper middle income = US$1,601 *per capita* or more.
Source: World Bank (1987).

Case study A

The impact of seasonality

The case study is set in a tropical environment where a wet season follows a dry season, and where cultivation is practised. Towards the end of the dry season, food becomes scarcer, less varied, and more expensive. The poorest people, who may be landless or have only small plots of land, experience food shortage more acutely than their less poor neighbours. Some migrate in search of work and food. Others undertake non-agricultural activities near their homes in which the returns are low. More work is involved in fetching water.

When the rains come, land must be prepared, and crops sown, transplanted, and weeded. If animals are used for ploughing, they are weak after the dry season. Delays in cultivation reduce yields. For those with land, food supplies depend on the ability to work or hire labour at this time. For those without land, work in the rains and at harvest often provides the highest wages of the year. This is the time of year when food is most needed for work, but it is also the hungry season, when food is in shortest supply and most expensive. It is, too, a sick season, when exposure to tropical diseases is at its greatest, when immunity is low, and when women are most likely to be in late pregnancy. So the rains bring crisis. Vulnerable to hunger, sickness, and incapacity, poor people are undernourished and lose weight. Seasonal stress drives them into debt and dependence. The knowledge that there will be future seasonal crises constrains them to keep on good terms with their patrons. They are thus seasonally forced into subordinate and dependent relationships in which they are open to exploitation. The poor are subordinated to the less poor and the weak to the strong. Stress is passed down to the weakest – women, children, old people, and the indigent. Sometimes this becomes an irreversible downward movement into deeper poverty as assets are mortgaged or sold without a hope of recovery. This is, then, a time when poor people are kept poor and a time when they become poorer.

With the harvest things improve. Grain prices are lower, a benefit to those who must buy food but a disadvantage to those small farmers who must sell their crops to repay debts or raise money for ceremonies, celebrations, and marriages, which take place after the harvest. Body weights recover. The dry season sets in. And then the cycle begins all over again.

(*Source*: Chambers *et al.* 1981:5)

reductions in the total number of rural people in Mexico, Bolivia, Brazil, Chile, Uruguay, and South Korea.

Defining the rural sector

The definition of 'rural', like that of 'urban', is the subject of much debate. In general the arguments over the rural sector mirror those presented in David Drakakis-Smith's book in this series (*The Third World City*). It may well be that too much attention is devoted to the need for universal definitions of urban and rural. In many instances the division is a highly artificial one which does not necessarily further our understanding, and there is a danger that by analysing the rural and urban sectors separately we overlook the processes common to both.

Urban centres generally operate as the transmitters of change to the rural areas. In recent years the degree of urban–rural interaction has increased dramatically. The development of communications and commercialization have resulted in the 'distinctiveness' of rural areas being broken down. Aspects of metropolitan life and world events now impinge directly on most Third World rural communities. High levels of temporary and permanent migration (see Graeme Hugo, *Population Movements and the Third World*, in this series) are major influences on bringing rural and urban areas closer together in terms of their characteristics.

Table 1.3 The changing role of agriculture, 1965–85 (%)

Agriculture's share of	Low income	Lower middle income	Upper middle income
GDP			
1965	41	29	15
1985	32	22	10
Change	-9	-7	-5
Export earnings			
1965	64	63	38
1985	31	29	16
Change	34	34	-22
Labour force			
1965	77	65	45
1985	72	55	29
Change	-5	-10	-10

Source: World Bank (1987).

Table 1.4 Changes in Third World export prices and the terms of trade (%)

	1965-73	1973-80	1981-6
(a) Third World export prices			
All exports	6·4	14·0	-14·9
Food	5·3	9·1	-10·7
Non-food agricultural	4·5	10·3	-31·4
(b) Terms of trade*			
All Third World	0·7	1·6	-7·1
Low Income	1·7	-2·5	0·7
Middle income	0·6	2·2	-8·3
Sub-Saharan Africa	0·3	13·4	-34·1

* The terms of trade are the ratio of export prices to import prices. A fall in the terms of trade means that the price of exports has fallen relative to the price of imports. Thus more will have to be exported to pay for the same volume of imports.

Source: World Bank (1987).

Agricultural production

Agriculture remains the main source of livelihood for the majority of Third World people. However, over the last twenty years its relative importance has generally declined. As indicated in table 1.3, the reduction in the significance of agriculture has been greatest with respect to export earnings. While for some countries this reflects the expansion of manufactured exports, it is primarily a result of falling international prices for most Third World agricultural commodities and a deterioration in the terms of trade (table 1.4).

Agriculture retains an importance as a source of employment out of all proportion to its contribution to export earnings or gross domestic product (GDP). In part this is due to the difficulty of valuing subsistence production for GDP calculations and the general under-recording of agricultural production. For most Third World countries the contribution of agriculture to GDP is probably underestimated. However, the tendency for agriculture's share of GDP to be significantly below its share of employment is also a reflection of the generally low productivity of the sector.

For most of the Third World agriculture remains the single most important sector of production (figure 1.3). For a large number of countries agricultural production levels and export earnings largely determine not only the livelihood of the majority of the population but also the fortunes of the economy as a whole. A number of Third World

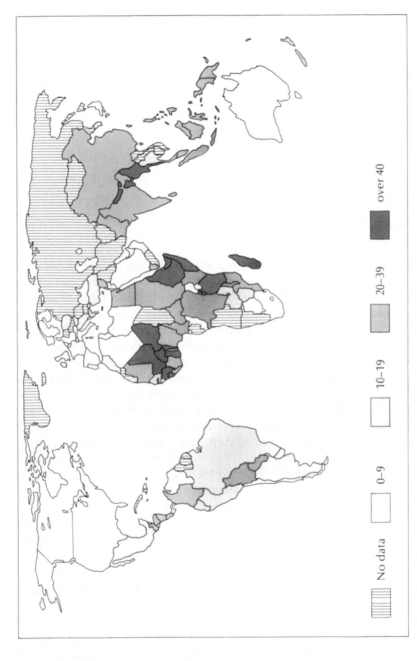

Figure 1.3 The percentage share of agriculture in GDP. *Source:* World Bank (1988)

No data

0–9

10–19

20–39

over 40

Table 1.5 Countries heavily dependent on one agricultural product for their export earnings, 1985

Country	Crop	Share of export earnings (%)
Burundi	Coffee	85
Colombia	Coffee	65
Cuba	Sugar	86
Ethiopia	Coffee	77
Ghana	Cocoa	75
Malawi	Tobacco	57
Seychelles	Oilseeds	65
Somalia	Livestock	87
Uganda	Coffee	86

Source: UNCTAD (1987).

countries remain dependent for their export earnings on one or two crops (table 1.5), thus greatly increasing their vulnerability to price or harvest variations. Between 1970 and 1984 Third World agricultural production expanded by 3·3 per cent a year, almost three times the rate for the developed countries. Despite continued high rates of population growth, this increase was sufficient to raise *per capita* production by 1·0 per cent a year. These figures hide very considerable regional and national variations. In South America between 1974–6 and 1982–4 total agricultural production grew by 26 per cent and *per capita* production by 5 per cent; in contrast in Africa, while production expanded by 14 per cent, *per capita* production fell by 10 per cent. Figure 1.4 indicates the very great variations between countries that lie behind even these regional contrasts.

Decline in production was heavily concentrated in Africa. Of the forty-seven African countries for which there are comparable data, twelve experienced a decline in total production and a further twenty-four in *per capita* production. Of the fifteen Third World countries with a decline in *per capita* production of 20 per cent or more, twelve were in Africa. The exceptions were Nicaragua, Trinidad, and Bolivia. Despite the comparatively rapid overall rates of growth of Third World production, the agricultural sectors have generally grown at a slower rate than the economies as a whole (table 1.6). Between 1965 and 1980 only in Liberia and Ghana did the growth of the agricultural sector exceed the rate of national growth. In contrast, between 1980 and 1985 for a whole range of middle-income economies agricultural growth rates exceeded national rates. This was a reflection not of the strength of the agricultural sector but rather of the weakness – in some cases the virtual collapse – of the other

Figure 1.4 Percentage change in *per capita* agricultural production, 1974–6 to 1982–4. *Source:* United Nations (1987)

No data

-45 to -20   -21 to 4   5-30   31-56   57-82

Table 1.6 Rates of growth of GDP and agriculture

(a) Rates of growth of GDP and agriculture, 1965–85 (% p.a.)

	GDP		Agriculture	
	1965–80	1980–5	1965–80	1980–5
Low income economies	4·8	7·3	2·7	6·0
Lower middle income economies	6·3	1·6	3·3	1·9
Upper middle income economies	6·6	1·7	3·7	2·3

(b) Countries where agricultural growth exceeded that of GDP, 1980–5

Argentina*	Mauritania
Brazil*	Mauritius
Central African Republic*	Mexico*
Chile*	Nicaragua*
Costa Rica*	Panama*
Honduras*	Peru*
Jamaica*	Ruanda*
Jordan	Somalia*
Liberia*	Trinidad*
Malawi	Venezuela*
Mali	Zimbabwe

*Indicates significant decline in the manufacturing sector

Source: World Bank (1987).

sectors of the economy, most notably manufacturing. Perhaps the most spectacular example is Brazil. Between 1965 and 1980 the Brazilian economy grew at an average annual rate of 9 per cent – manufacturing expanding at 10 per cent and agriculture at 4·7. In contrast, between 1980 and 1985 economic growth averaged only 1·3 per cent – manufacturing growing at only 0·3 per cent and agriculture at 3·0 per cent.

Food production

In looking at Third World agriculture it is important to separate domestic food production from raw material and export-oriented crops. For many Third World countries export crops occupy the best agricultural land, absorbing the majority of modern agricultural inputs, investment, and agricultural development expenditure. Figure 1.5 gives some indication of the proportion of land devoted to these crops. The variation is considerable, ranging from over 30 per cent in such countries as Malaysia, Colombia, and Sri Lanka, to less than 1 per cent in Bolivia. In countries heavily dependent on agricultural export earnings, falling commodity

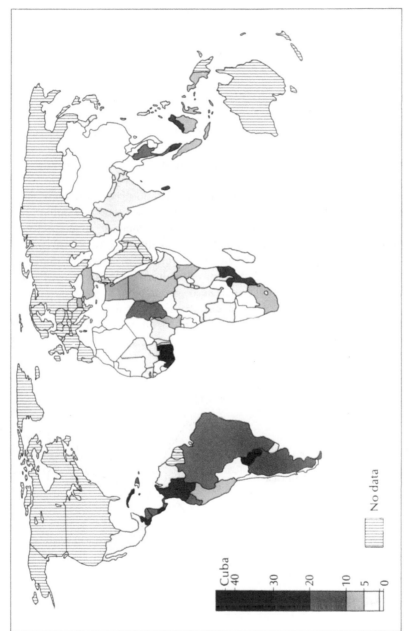

Figure 1.5 Percentage of cultivated land devoted to export crops. *Source:* Open University (1982: 38)

prices and worsening terms of trade (see table 1.4) for Third World primary production have caused a switch of resources away from domestic food production.

While the pattern of change in *per capita* food production is very similar to that of agricultural production, there are some notable exceptions. Again there are striking regional contrasts (figure 1.6), with sub-Saharan Africa experiencing a very steep decline. Between 1979 and 1985 of the twelve countries with declining *per capita* food production, eight were in sub-Saharan Africa.

For the most of the Third World, *per capita* food production increased between 1974 and 1984, but cereal imports also rose, from 40·9 million to 71·3 million metric tons. It is important to put this increase in perspective. In both years imports represented only 0·1 per cent of Third World cereal production. However, it is clear that for a number of countries imports have become an increasingly important element in the food supply. Food needs in the Third World have increased far more than the figures for imports suggest. Imports reflect a country's ability and willingness to purchase cereals: these may be very different from the people's need for additional supplies.

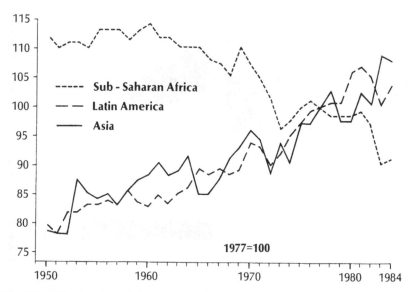

Figure 1.6 Regional trends in *per capita* food production. *Source:* Guthrie (1986: 93)

The reasons for the divergent trends in Third World *per capita* food production are complex. Explanations advanced have included: adverse environmental conditions, unsuitable agricultural practices, over-population, underpopulation (in parts of West Africa) and neglect of the agricultural sector. For many writers these items, singly or in various combinations, account for the 'failure of Third World agriculture'. Others see these 'explanations' as symptoms of the problem, whose causes lie in the evolution and operation of the international economy.

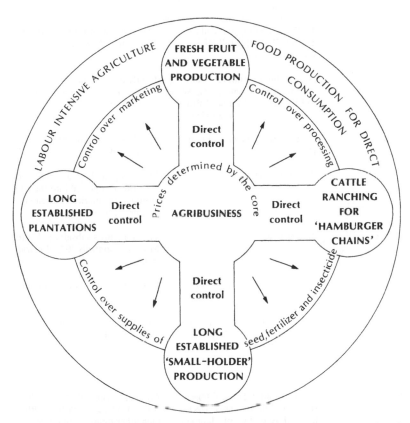

Figure 1.7 World agriculture. A simplified representation of the relationship of Third World agriculture to the capital-intensive, market-orientated agriculture of the developed countries, which is increasingly dominated by agribusiness. *Source*: partly based on Bradley (1986)

Third World and Developed World agriculture

The problems of Third World agriculture cannot be understood in isolation. Figure 1.7 summarizes the interrelationships of Third World and Developed World agriculture. It is important to understand that Developed World agriculture is not merely a more sophisticated version of Third World cultivation. The systems of production are fundamentally

Table 1.7 Multinational control of major Third World crops

Crop	Companies	Comments
Cocoa	Cadbury–Schweppes, Gill & Duffus, Rowntree (all UK), Nestlé (Swiss)	These four companies control 60–80% of world cocoa sales
Tea	Brooke Bond, Unilever, Cadbury –Schweppes, Allied Lyons, Nestlé (all European); Standard Brands, Kellogg, Coca-Cola (all US)	Jointly hold about 90% of tea marketed in Western Europe and North America
Coffee	Nestlé (Swiss), General Foods (US)	Jointly hold around 20% of the world market
Sugar	Tate & Lyle (UK)	This company buys about 95% of cane sugar imported into the EEC, although the market is threatened by European beet sugar
Molasses	Tate & Lyle	The company's subsidiary, United Molasses, controls 40% of world trade
Palm oil	Unilever (UK-Dutch), Lesieur (French)	Unilever dominates trade in palm oil
Tobacco	BAT (UK), R. J. Reynolds (US), Philip Morris (US), Imperial Group (UK), American Brands (US), Rothmans (UK-South African)	Together control between 89% and 95% of world leaf tobacco trade
Cotton	Velkart, Cargill (US), Bunge (Dutch), Ralli Brothers (UK), Soga Shosho (Japan), Bambax, Blanchard	Are major transnationals in the cotton trade, and together with nine other multi-commodity trading groups they dominate 85–90% of world cotton trade

Source: Based on Dinham and Hines (1983).

different. Indeed, much Developed World agriculture has more in common with manufacturing than with peasant agriculture in the Third World.

Developed World agriculture may be characterized as capital-intensive and market-orientated. Production is for sale in the same way as with manufactured goods. Crop and livestock complexes are largely responses to market forces. Land use, for example, is highly responsive to changing price levels. In consequence some writers have labelled Developed World agriculture as 'agribusiness'.

The developed countries exert considerable control over world agriculture (table 1.7). The price levels of most tropical and sub-tropical crops, such as rubber, tea, coffee, or cocoa, have always been determined by the Developed World consumers. However, increasingly the agricultural policies of the United States and the European Economic Community are governing price levels. These policies are frequently responses to the influential 'farm lobbies'.

Since 1978 the EEC has restricted imports of Third World cassava and maize in order to protect the interests of southern European producers of low-grade grain. On a wider scale, since the early 1970s, world wheat prices have moved largely in parallel with the United States' domestic price support policy. Harvest conditions and level of demand have had minimal impact on price.

In contrast, Third World countries have little influence over price levels. Third World agriculture may be typified as the small-scale, labour-intensive production of food for direct consumption. While production for Developed World markets has been long established, it has been largely separate from food production for local consumption.

An important element in the incorporation of Third World agriculture into the international market has been the emergence of multinational agribusiness. Corporations exert increasing control over Third World agriculture through marketing and the supply of inputs, such as seeds and fertilizers, as well as by direct control over the production of crops. These developments are discussed in more detail in chapter 3.

Key ideas

1 The majority of Third World people live in rural areas. But, whilst agriculture remains central to the life of most rural communities, a wide range of other activities also occur.
2 In most of the Third World the distinction between rural and urban communities is being rapidly broken down.

3 Conditions in rural areas often vary greatly between seasons and from one year to the next.
4 Agricultural production is usually underestimated in national statistics. However, even allowing for this, the sector has generally become less important, growing at a slower rate than the manufacturing and service sectors.
5 Third World agriculture is not just a less sophisticated version of that practised in the Developed World. Much of it is based on a fundamentally different system of production. Third World agriculture does not operate in isolation. It is very much part of an international economy largely controlled by Developed World interests.

2
Traditional rural production systems

The term 'traditional' used in this chapter refers to systems of production that owe little or nothing to incorporation into the world economy. Few such systems now survive. The process of incorporation has either eliminated or greatly modified them. However, Third World agriculture still contains much that could be described as 'traditional', or perhaps more accurately 'pre-capitalist'. An examination of the nature of these systems, their complexity, underlying principles, and capacity for change is essential if the agriculture of the contemporary Third World is to be understood.

In pre-colonial Africa, Asia, and Latin America agricultural production was mainly for direct consumption by farm households and the immediate community – a very similar situation to that of pre-industrial Europe. However, surplus production and trade in agricultural produce were important over large areas. The highly developed urban cultures of Asia, for example, were dependent on the regular production of a large agricultural surplus. In the seventeenth century rice was regularly shipped from parts of South East Asia to China, while in the late fifteenth century it is estimated that 30 per cent of Asian spice production (nutmeg, cloves, pepper, cinnamon) reached Europe. In the eighteenth and nineteenth centuries the large volume of goods traded from India and China was produced by sophisticated pre-capitalist production systems.

Traditional agricultural strategies

Traditional systems of production and indeed much of contemporary Third World agriculture remain geared to production for direct household consumption. The survival of the community depends on the production of at least the subsistence minimum each year. For members of traditional communities a poor crop may mean food shortages, starvation, illness, and even death. In order to survive households may have to dispose of, or mortgage, such essential assets as land and livestock. In extreme situations next year's seed may have to be eaten. Thus the price of surviving a poor harvest may be a permanent reduction in the household's ability to cope even with normal conditions.

The reliability of production levels is likely to be more important to farmers than either yield or return to labour. A crop that produces a low but reliable yield may be more acceptable than one that produces a higher but more vulnerable output. This is illustrated in figure 2.1. The traditional variety of rice produces a low but consistent yield. Only once in the thirty-year period does production fall below the subsistence miminum. In contrast the improved variety fails to provide for subsistence needs in eight of the thirty years. While the total output of the improved variety is much greater over the 30-year period the farmer may well not have survived to benefit. Only in the unlikely case of the farmer having the facilities to store rice over a long period could there be real gain from the improved variety.

Where environmental conditions are uncertain very complex strategies may be developed to compensate (case study B). These are designed to ensure the subsistence minimum in all but the very worst years. Where they fail to do so there are likely to be additional 'safety nets' before assets are threatened. These may include: obtaining alternative poorer foods – for example, cassava could be substituted for rice (in extreme conditions various wild 'famine' foods may be gathered); seeking cash income or payment in kind from petty trading, handicrafts, or labouring; or temporary migration. Such elements are often permanently incorporated into household survival strategies.

In addition there are likely to be a variety of informal mechanisms which support poorer households, providing aid in periods of crisis. These are likely to be reciprocal and involve labour, necessities, and agricultural inputs, such as seed. Such structures tend to break down very rapidly with the development of commercial production (see chapter 3).

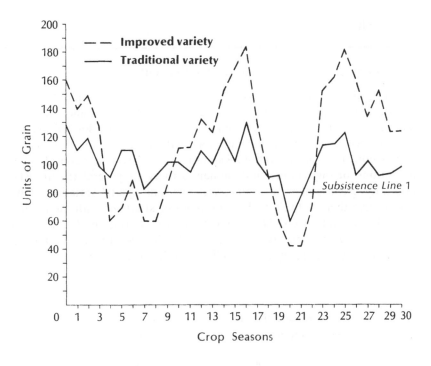

YEARS BELOW SUBSISTENCE

Variety	Number	Percentage
Traditional	1	3.3
Improved	8	26.6

Notes   *1   300 Kg / per capita of unmilled rice*

Figure 2.1 Rice yields and the subsistence farmer. *Source:* Scott (1976: 16)

Change in traditional agriculture

How and why does change take place in traditional agricultural systems? Certainly a vital ingredient is the introduction of more productive methods of cultivation and crops. Many studies have shown that knowledge of these techniques, perhaps obtained by contact with communities that practise them, can lie dormant. It is not always clear why it is not immediately put into practice.

Case study B

Rain-fed rice cultivation in north-east Thailand

The agriculture of north-eastern Thailand is dominated by the cultivation of glutinous or 'sticky' wet rice. This rice is the staple subsistence crop of a large part of Burma, Thailand, and Laos, but has only a limited market outside those areas. A relatively even annual temperature of 28°C places no real constraint on cultivation. Rainfall, in contrast, places considerable limitations on agriculture and human activity in general (figure B1), primarily because of wide annual, seasonal, and spatial variations. One year in five there is likely to be serious water shortage, with the annual rainfall total well below the minimum for rice cultivation. The seasonal pattern, too, is far from stable, with considerable variation in the starting date of the wet season, even over quite short distances.

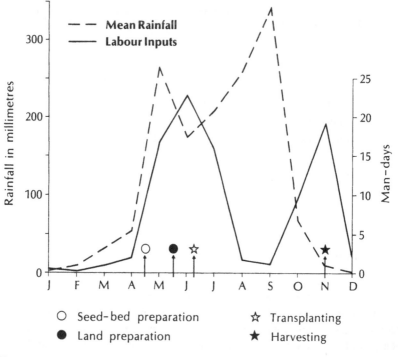

Figure B1 The rice year and the rainfall pattern

Case study B *(continued)*

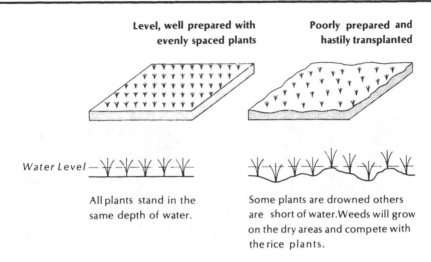

Figure B2 Different degrees of land preparation

These factors combine with topography to produce local variations in water supply and, therefore, in both the date of land preparation and the amount of land that can be planted. Indeed, one of the most distinctive features of the region's agriculture is the very large annual fluctuation in the area planted. Late rains reduce the time available for land preparation. Farmers often cannot plant all their land, and some will be less well ploughed and levelled. In wet rice cultivation poor land preparation reduces yields (figure B2). In addition, since the rice varieties grown flower and mature in response to a particular length of daylight, late planting reduces the growing period and the yield.

Population growth and the expanding need for surplus products have resulted in steady growth of the cultivated area. There has been little of the intensification more usually associated with Asian paddy. Under north-eastern rainfall conditions concentration on a small holding would be extremely risky. Extensive cultivation enables the risk to be spread. Not only do land holdings in the north-east tend of necessity to be large and not intensively cultivated, they also show a marked tendency towards fragmentation. The typical farm consists of a series of discontinuous holdings, as illustrated by farm D in figure B3.

Case study B *(continued)*

Plate B1 Unintensive rice cultivation in north-east Thailand. These fields have been cultivated for 300 years. Note the many large trees and termite mounds left in the fields

Traditional inheritance practices preclude subdivision of the holding, with the land passing intact to a son-in-law through a daughter, normally the youngest. The other siblings acquire land through marriage or by clearing new land. Some of the fragmentation may be explained by the clearance of land at different dates to enlarge the family holding, but this affords only a partial explanation. A more basic reason is the nature of the environmental conditions and paddy cultivation practices which have developed in response to them.

The unreliable environmental conditions of the north-east result in few farmers willingly holding one plot of homogeneous paddy land. In wet years, low-lying land is either inundated and unplantable or subject to extensive crop loss. Similarly, in dry years high land may be too dry to plant or the crop may be seriously damaged by drought. Farmers attempt to spread the risk by holding land in a variety of situations.

Case study B *(continued)*

Figure B3 Land use and land holding in a part of Ban Tum, a village in Kalasin province

The ideal pattern of land holding is shown in Figure B4. Plots are generally elongated, at right angles to the contours, thus achieving several water supply situations within the confines of a single plot (for example, farm A on figure B3). Ideally these elongated plots encompass a variety of land types, from low flood-prone land suitable for seedbeds early in the year, through a variety of paddy land, to upland and even forest, but in the

Case study B *(continued)*

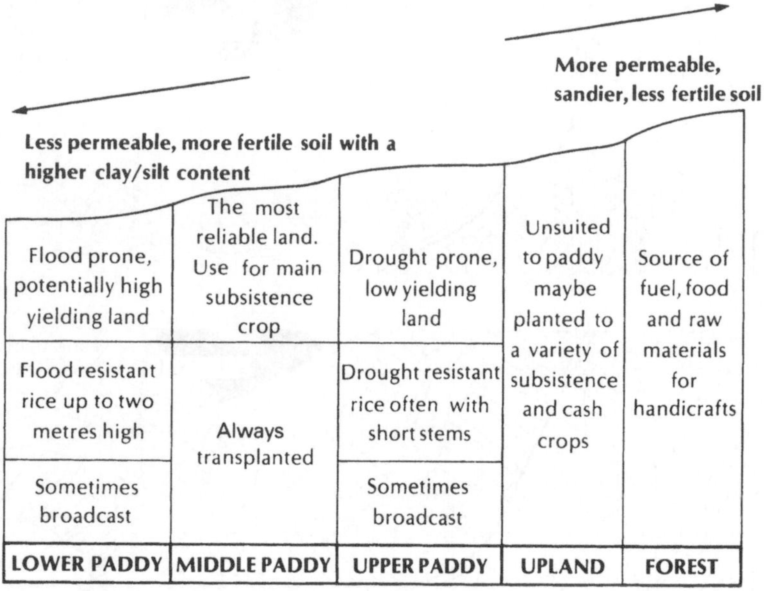

More permeable,
sandier, less fertile soil

Less permeable, more fertile soil with a
higher clay/silt content

Flood prone, potentially high yielding land	The most reliable land. Use for main subsistence crop	Drought prone, low yielding land	Unsuited to paddy maybe planted to a variety of subsistence and cash crops	Source of fuel, food and raw materials for handicrafts
Flood resistant rice up to two metres high	Always transplanted	Drought resistant rice often with short stems		
Sometimes broadcast		Sometimes broadcast		
LOWER PADDY	**MIDDLE PADDY**	**UPPER PADDY**	**UPLAND**	**FOREST**

Figure B4 Idealized land-use cross-section

village shown in figure B3 most farmers achieve a holding with a full range
of land types by acquiring a number of scattered plots, for example farm D.
The variety of land in any one holding also makes the peaks of labour

Explanations that involve cultural preference, innate conservatism, or
the long period of assimilation necessary before new methods or crops are
adopted may contain elements of truth but they do not provide a complete
answer. Increased productivity, which in modern capitalist society is
usually accepted as a desirable thing in itself, may have little attraction for
traditional cultivators. There has to be a need for increased production
which the existing practices cannot easily or safely meet. The decision to
innovate will involve the balancing of need against the increase in labour
input, the level of risk, or the sustainability of the increased output. Only
under great pressure are farmers willing to adopt more risky methods. In

Case study B *(continued)*

---

requirement rather less sharp than would be the case with a homogeneous holding, so that a larger area can be cultivated with the same amount of labour.

The effectiveness of holding a variety of land types as an insurance against adverse environmental conditions and to stagger labour peaks is enhanced by the planting of a variety of rice types, with different optimum water needs, stalk lengths, and maturation periods. The planting of these different varieties on different land is only part of a complex of detailed knowledge that each farmer has concerning the variety of the soil, water supply, and topography of his holding as well as of environmental probabilities. The introduction of improved paddy strains with standard characteristics severely limits these practices, so that the improved varieties have been most widely adopted in the areas of more reliable production conditions.

The least reliable land is usually cultivated with a lower degree of intensity, and in some cases broadcasting replaces transplanting, an exercise that reduces labour input by as much as a third. In areas where a commercial element has developed, non-glutinous rice for sale is likely to be grown on the least reliable land, for example the flood-prone land near the lake on Figure B3 (b), the most reliable being reserved for the household's own glutinous rice.

This long established system is coming under increasing pressure as the supply of uncleared land suited to paddy rice cultivation is becoming exhausted.

(Adapted from Dixon 1978)

---

the traditional system such pressure for increased output usually comes from the need to pay taxes or other levies, produce for exchange, or feed a growing population. During the colonial period a variety of methods were used to increase farmer's needs for surplus production in order to develop cash cropping (see chapter 3).

For most agricultural systems long-term population growth was probably the single most important factor in stimulating an increase in production. Increased numbers not only made increased output necessary but also provided the labour to make it possible. For many crops substantial long-term increases in yield can be achieved by increasing the labour input. This is particularly the case for wet rice.

For many writers population growth is the key not only to the intensification of the existing cultivation methods but also to the transition to new, more intensive systems. This process may also include diversification into handicraft production and trading. Historically these developments usually reflected the ability of the agricultural system to support such activities. However, the growth of non-agricultural activities can also result from the inability of agriculture to support the community. For much of the contemporary Third World this is increasingly the case (see chapter 3).

Shifting cultivation and intensification

Shifting systems of cultivation were once much more widespread than they are today but they remain of major importance in much of tropical and sub-tropical Africa, South America, and to a lesser extent Asia. The essential feature is the temporary cultivation of land followed by a lengthy period of fallow during which vegetation cover and soil fertility are re-established.

Plate 2.1 Shifting cultivation. The field has a partially cleared appearance. Note the large tree trunks left lying across the slope to reduce soil loss

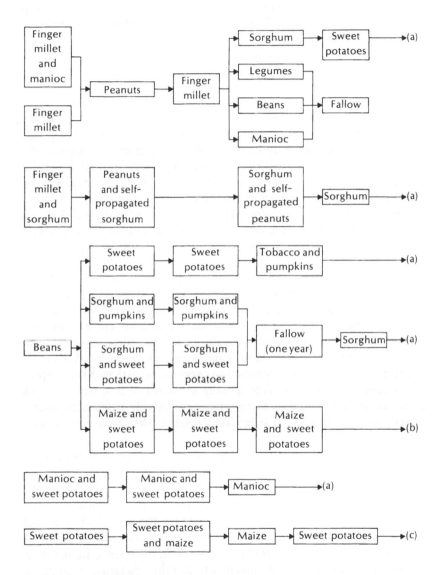

Figure 2.2 Shifting cultivation crop sequence and inter-planting in Zimbabwe. (a) No information given on what follows. (b) Maize and sweet potatoes planted year after year until the soil is exhausted, then sweet potatoes alone as the land crop. (c) Maize and pumpkins sown so long as the soil is fertile, then tobacco or sweet potatoes. *Source:* Ruthenberg (1971: 32)

There is great variation in length of cultivation and fallow as well as in crop complexes. These reflect environmental conditions, population density, and cultural preference. In sub-Saharan Africa a very broad distinction may be drawn between the sub-tropical 'bush-fallow' and the tropical 'forest-fallow'. In the former, two to four years of cultivation of single crops such as maize, millet, sorghum, and peanuts is followed by six to twelve, or even more, years of fallow. The tropical forest system involves the cultivation for only one or two years of a wide range of crops which gives way to rather longer periods of fallow. The local and regional variations in these systems are considerable. The shifting cultivation of the tropical forest areas in particular has the apperance of chaotic partial cultivation (plate 2.1). Such systems were widely condemned by agronomists, particularly during the colonial period. Shifting systems were described as 'inefficient' and often 'damaging to the environment'. Today it is generally agreed that these systems are a logical and ecologically sound way of cultivating infertile tropical and sub-tropical soils, provided that the fallow period is long enough.

The complex sequence of inter-planting typical of many systems (figure 2.2) has a number of important functions. First, it produces a variety of crops for food, handicrafts, and exchange through the crop year. Second, it spreads labour inputs. Third, it maintains vegetation cover to prevent soil damage and in some cases provide shade and reduce moisture loss. Shifting systems are usually associated with subsistence production of traditional crops. However, communities dependent on them have adopted a wide range of new crops. Additionally a number of field and tree cash crops are produced under shifting regimes, notably cassava in Brazil. Indeed, contrary to popular belief, such systems often have considerable capacity for change.

Communities dependent on shifting cultivation respond to population increase or increased need for surplus production by expanding the area cultivated. If land resources are fixed or even reduced (for example, by the creation of forestry reserves) the fallow period will be shortened and land cropped more frequently. Figure 2.3 illustrates the impact of increased population on a shifting system where the land area is fixed. The period of unnecessary fallow in (a) represents the 'slack' in the system. Elimination of this produces in (b) the maximum carrying capacity. Continued population increase and reduction of the fallow period produce the disastrous situation illustrated in (c) and discussed at greater length in chapter 4.

Before the downward spiral is established the availability of additional

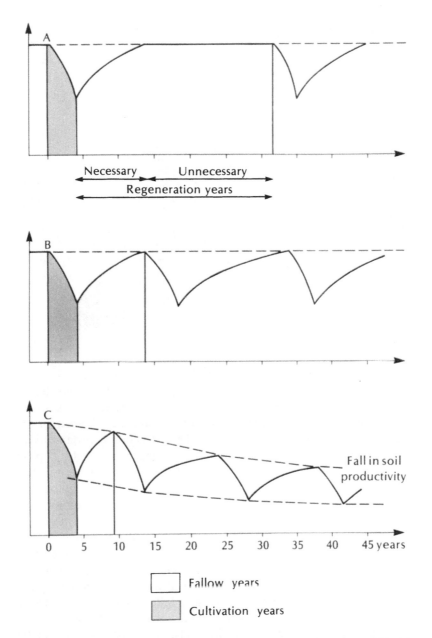

Figure 2.3 Intensification of shifting cultivation. *Source:* Ruthenberg (1971: 47)

Plate 2.2 Established wet rice landscape

labour may enable the transition to a more productive system of cultivation to be made. In South East Asia, since the mid-nineteenth century, large areas have been converted from shifting cultivation to sedentary wet rice. It is estimated that forty persons per kilometre is the 'trigger density' for much of the region. The transition necessitates very considerable labour input to create the wet rice landscape depicted in plate 2.2. The large labour input and the time necessary to get the wet rice system into operation at first results in the two systems operating together (plate 2.3). Shifting cultivation with supplementary wet rice on the land most easily converted then gives way to a predominance of wet rice as the more difficult land is developed. In Java between 1850 and 1900 shifting cultivation was largely eliminated. The intensive wet rice cultivation that replaced it was able to support an increase in population from 9·6 million in 1850 to 28·4 million in 1900. By the end of the nineteenth century the intensification of wet rice was beginning to falter, and *per capita* paddy production fell. The response was to plant dry-season food crops – maize, cassava, and sweet potatoes – in the rice fields. This practice spread rapidly between 1900 and 1930, maintaining *per capita* food production at, or

Plate 2.3 A pocket of terraced wet rice cultivation in an area still dominated by shifting cultivation

near, the nineteenth-century level.

Jave illustrates the ability of traditional systems to intensify and develop more productive methods of cultivation and adopt new crops. However, in the face of the restrictions placed on the Javanese economy by the Dutch colonial structure, intensification became increasingly precarious and rural Java now has some of the deepest and most widespread poverty in Asia.

Key ideas

1 Traditional rural societies have evolved agricultural systems in which limited technology is applied to enable the needs of subsistence and limited exchange to be sustained over long periods of time.
2 Traditional systems are capable of adapting to the growth of population, the introduction of new crops and techniques, and to a degree of commercialization.
3 There are limits to the ability of traditional systems to cope with rapid externally induced change.

3
Rapid rural change: traditional systems under pressure

The traditional agricultural systems described in chapter 2 have come under increasing pressure as they have been incorporated into the world economy. This process has centred on the 'commodification' of agricultural products. What this means is that the value of crops, as determined by their *localized* use for food or raw materials, is replaced by a value determined by market forces. Once this new (capitalist) form of production is established, those dependent on agriculture lose all control over the determination of the value of their products. The process of incorporation into the capitalist economy thus introduces a new form of production, as well as new crop complexes and methods of cultivation.

Third World crops

It is often forgotten that a large number of the crops that we consider the 'natural' produce of particular Third World countries were in fact introduced by the European powers. A very general summary of this global process of 'transplanting' is contained in figure 3.1. It is important to appreciate that a wide variety of food crops for local production have been introduced, as well as the more obvious export crops. During the seventeenth century the Portuguese brought South American maize, cassava, and chillies to many parts of Asia. In the same period, maize, peanuts, cassava, and Asian rice were introduced into Africa.

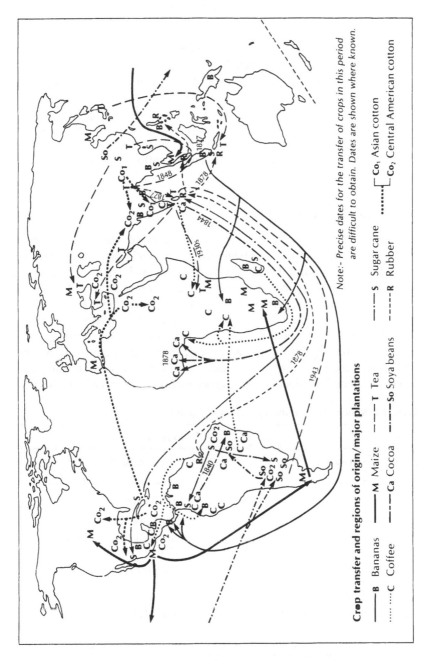

Crop transfer and regions of origin/major plantations

B Bananas   M Maize   T Tea

C Coffee   Ca Cocoa   So Soya beans

S Sugar cane   Co₁ Asian cotton

R Rubber   Co₂ Central American cotton

Note:- Precise dates for the transfer of crops in this period are difficult to obtain. Dates are shown where known.

Figure 3.1 Transfer of major crops. *Source:* Open University (1982: 29)

The introduction of commercial crops during the nineteenth and early twentieth centuries often reflected colonial rivalries, with the dominant centres of production changing several times. Coffee, which is indigenous to Ethiopia, has successively been produced in Sri Lanka, Java, Brazil, and now increasingly West and East Africa. Many colonies underwent a cycle of monoculture, with changes in the crop grown resulting from colonial competition reinforced by the susceptibility of intensive monoculture to disease. In Sri Lanka the sequence was cocoa, coffee, chinchona (quinine), rubber, and tea.

The process of incorporation

In general the process of incorporation involved the breakdown of localized subsistence production and the creation of a dependence on cash income from crop sales. Essentially the farmers involved were adopting a much more 'risky' agricultural strategy (see chapter 2). For farmers to replace a subsistence crop with a cash crop they have to accept an increased and uncertain level of risk. A successful subsistence harvest normally means that there will be sufficient food for the community. In contrast a 'bumper' cash-crop harvest carries no such guarantee. A large crop may result in low prices and insufficient income to purchase necessities. The introduction of production for the market is likely to be

Table 3.1 Major forms of incorporation of agriculture

Form	Example
Plantations and estates	Caribbean, Brazil
White settlements	Kenya, Rhodesia, South Africa
Taxation, rent, and other levies in specified crops	Dutch East Indies
Taxation, rent, and other cash levies	West Africa
Taxation, rent, and other levies in labour	Dutch East Indies, much of Latin America
Creation of 'cash needs' through the establishment of state mangolies over products previously domestically produced	French Indochina
Creation of new cash needs to purchase imported goods	Burma, Thailand
Restriction of land area for traditional agriculture, then 'squeezing' the community into work in the commercial sector	East and West Africa

Table 3.2 Comparative effect of proportional and fixed taxation on traditional cultivators

	Net yield (baskets)					
	50	40	35	30	25	20
Proportional tax						
20%	10	8	7	6	5	4
Remaining	40	32	28	24	20	16
Surplus (deficit) over subsistence figure of 25						
baskets	+15	+7	+3	–1	–5	–9
Fixed tax						
Eight baskets (20% of average yield of 40)	8	8	8	8	8	8
Remaining	42	32	27	22	17	12
Surplus (deficit) over subsistence figure of 25						
baskets	+17	+7	+2	–3	–8	–13

Source: Scott (1976: 68).

easier where the existing subsistence crop is the potential cash crop, or the new fits into the existing farming system and does not compete for land or labour. Table 3.1 summarizes the main ways in which cash crop production was introduced. In practice many of these operated in combination.

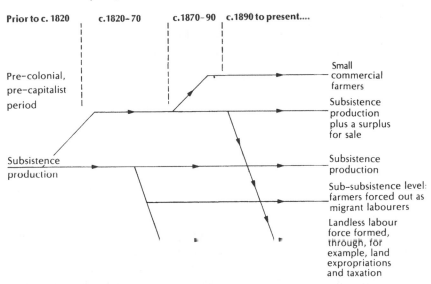

Figure 3.2 The differentiation of rural communities in southern Africa. *Source:* based on Webster (1986)

Table 3.3 Indian agrarian structure, c. 1960

Landlords
  Large-scale absentee owners of land, often with land in more than one village. Most common in West Bengal, eastern Uttar Pradesh, Bihar, Orissa, parts of Tamil Nadu and Andhra Pradesh.

  Smaller, normally resident, owners, usually holding land in one village. Most common in Maharashtra, Gujarat, parts of Tamil Nadu, Madhya Pradesh, Assam, and Uttar Pradesh.
  Some used labour from poor and landless households to cultivate part of their holdings.

Rich farmers
  Part owners, part tenants; sold most of their produce and often hired labour. Generally they worked their holdings less intensively than the middle-income and poor groups.

Middle income farmers
  Part owners, part tenants; owned a larger proportion of their land than poor peasants but a smaller proportion than the rich. Sold some of their production and often hired labour at peak periods.

Poor farmers
  More likely to be tenants, often paying rent in kind or operating as sharecroppers. Produce often sold not because it was surplus to consumption needs but because of pressing cash needs. Frequently worked for large land operators.

Landless
  Permanent: tied for a year or longer.

  Casual: hired only for specific tasks – weeding, harvesting, threshing.

  By 1951 15 per cent of agricultural households were landless.

Source: compiled from Byres and Crow (1983).

In the pre-colonial period taxes and other levies on the rural communities were generally adjusted to the varying levels of production. The imposition of colonial rule invariably resulted in these claims becoming fixed, regardless of harvest levels. Table 3.2 illustrates the differential impact of fixed and proportional levies. From the limited information available it seems probable that traditional tax demands were often rather lower than the 20 per cent used in this example. Additionally there is evidence of a reduction in the proportions in years of poor harvests. A similar change took place with respect to land rent. Under the traditional system, where rent was payable, it was likely to vary with production. Once the rent was fixed the burden of 'risk' was shifted entirely to the tenant cultivator and away from the landlord.

The establishment of production for the market, fixed cash, rent, taxes, and wage labour, all set in motion a process of differentiation (figure 3.2). Land could be sold or taken by landlords and moneylenders in default of payment. Small farmers in some cases lost their land and became landless labourers, while large farmers, merchants, moneylenders, and landlords acquired land. Table 3.3 illustrates the complex structure of rural communities that had emerged in India by 1960.

In general, colonial agricultural policy was aimed at facilitating the development of commercial production. However, in a number of cases subsistence production was seen as the 'ideal' for all or part of the indigenous population. The Malays were labelled 'natural subsistence rice cultivators' and effectively excluded from other activities. Similarly in Kenya areas were set aside for 'indigenous subsistence cultivation'. However, the process of commercialization was a far from peaceful one. The implementation of the policies summarized in table 3.1 met with widespread and often protracted resistance. During the 1930s major rural 'rebellions' against colonial policy occurred in French Indochina, the Dutch East Indies, Burma, India, and Jamaica.

Direct incorporation: plantations and settler economies

The most direct, and chronologically the earliest, incorporation of colonial land and labour into the world market was the establishment of plantation economies. These developed from the early seventeenth century to produce a range of tropical and sub-tropical crops, using slave labour. During the eighteenth and nineteenth centuries over nine million Africans were transported to the Americas, principally to produce sugar. This gave way to the employment of wage labour in some cases even before the abolition of slavery, because it was more productive and more profitable.

Many of the plantation companies became the core of present-day multinational agribusiness corporations. Between 1911 and 1930 Lever Brothers (later Unilever) established oil palm plantations in what are now Ghana, Zaire, Nigeria, Gabon, and the Cameroons. During the 1920s Brooke Bond developed tea estates in present-day Kenya, Sri Lanka, Pakistan, and India. Most spectacularly, in 1926 Firestone moved into Liberia, purchasing over 30,000 ha at US$1 per hectare, to establish the world's largest rubber plantation and earn the country the label 'the Firestone Republic'.

The end of the colonial period resulted in a general reduction of foreign ownership of plantations. Some were taken over by local concerns and

Case study C

Vegetables for the European market

Pan African Vegetable Products is a company set up by the Bruecker Werker company of West Germany, with part of the equity held by Barclays Overseas Development Company, the Industrial Co-operative Development Corporation (ICDC) of Kenya, the Industrial Development Bank, and the Safida Investment Company of Switzerland. The company grows carrots, turnips, beans, leeks, potatoes, dill, red and savoy cabbages, capsicums, swedes, and celery. The vegetables are dehydrated and sent to West Germany and other European countries at the rate of 36,000 metric tons a year. The company itself owns one farm and leases another, cultivating over 800 acres of irrigated crops, or half the vegetables required. The other half is supplied by more than 5,000 growers. These growers depend totally on the company: it approves and provides seeds, pesticides, and herbicides to listed growers on a credit basis, against the crops being grown for the factory. Over 90 per cent of the factory's production is exported, and it runs all year round, working six days a week, twenty-four hours a day, employing between 450 and 650 people, depending on the season.

In 1972 Bud Antle Inc., a large California-based food conglomerate (taken over in 1978 by Castle & Cooke, an American transnational with plantations in Latin America and Asia), formed a joint enterprise with the Senegalese government. The subsidiary, Bud Senegal, is an affiliate of the House of Bud in Brussels. Bud Senegal grew vegetables, using a virtually labour-free drip irrigation system, with plastic tubes continuously supplying water to each plant individually, tapping the vast reserves of water just below Senegal's dry soil. Three times a week, from early December until May, a DC-10 cargo jet takes off from Senegal loaded with green beans, melons, tomatoes, aubergines, strawberries, and paprika. The destinations are Amsterdam, Paris, and Stockholm. The vegetables are not marketed locally, but in any event few Senegalese have enough money to buy them.

Local people gained few jobs from the project, and in laying out its 450 ha plantation Bud uprooted the idigenous baobab trees which were an important village resource, having previously provided local families with rope, planting materials, fuel, and wind erosion protection. Other villagers fared even worse: the government provided police to evict them from their land to make way for Bud's plantation.

(*Source:* Dinham and Hines, 1983: 31–7)

others broken up into smallholder operations. However, foreign-owned plantations and estates remain significant producers of a number of Third World crops. As was noted in chapter 1, multinational agribusinesses have moved increasingly into control over marketing, processing, the supply of seeds, fertilizers, machinery, and other inputs. Additionally they have in recent years been attracted to Third World locations to take advantage of cheap land and labour to produce a range of new products – meat, fruit, vegetables, and fresh flowers – for Developed World markets (see case study C).

Smallholder developments frequently followed plantation development, in many cases acting as an 'expansion tank' for variation in demand, as for example in the Malaysian rubber-growing areas. Private ownership of land and sometimes land colonization schemes were important in smallholder development. In many instances small-scale cash-crop production linked with subsistence production developed under the stimulus of the policies listed in table 3.1. It was often far more profitable and less risky for companies to concentrate on collecting, processing, and supplying inputs such as seeds and fertilizers than to engage directly in production.

A second 'direct' method of incorporation was the establishment of 'white settler economies', such as those in South Africa, Kenya, Zimbabwe, and Algeria. The establishment of large European farming communities usually involved the resettlement of the indigenous population. The 'native reserves' often comprised land of low fertility which rapidly became overcrowded. Such areas frequently served as a convenient source of cheap labour (see case study D). Enormous European estates of prime land were established, often at little cost. For example, during 1903 in the Njora district of Kenya the following estates were obtained at less than one US cent per hectare: Lord Delamere (141,643 ha), Lord Francis Scott (40,469 ha), East African Estates (141,643 ha), and East African Syndicate (40,469 ha). The division between the indigenous and large-estate agriculture established in the late nineteenth and early twentieth centuries remains a major feature of many of these countries.

Traditional systems under pressure

Chapter 2 examined the ability of traditional systems to adopt new crops and more productive methods, and adjust to increased population and the need for surplus production. The process of incorporation discussed above generally undermines the traditional system, making agricultural communities more vulnerable to both economic and environmental

Case study D

Kenya: land under pressure

In Kenya the traditional agricultural system is widely reported as being under serious pressure. Soil erosion, deforestation, overgrazing, and falling *per capita* land holding are depicted as the result of very rapid population growth (3·5 per cent a year) and inappropriate cultivation technqiues.

There is a general relationship between the agro-ecological potential of the land (figure D1) and population density. Some 60 per cent of the population live in the high potential areas, 30 per cent in the medium to low, and 10 per cent in the arid areas. The most serious problem of environmental degradation occurs in the marginal and semi-arid areas. Signs of soil erosion are widespread, particularly on slopes of more than 15° (plate D1). In these areas desperate attempts to produce subsistence crops have resulted in maize cultivation being pushed up-slope, where it can be sustained for only one or two years.

Plate D1 Slopes badly eroded as a result of maize cultivation

Case study D *(continued)*

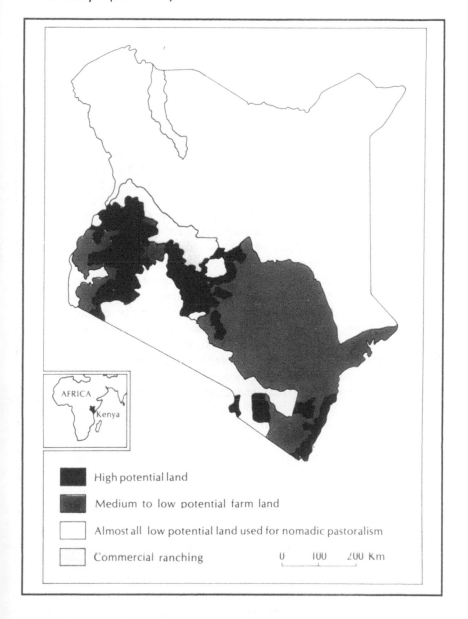

Figure D1   The main agro-ecological zones of Kenya. *Source*: Hunt (1984: 15)

Case study D *(continued)*

Reliable data on the situation in the subsistence sector of rural Kenya are few, but large-scale environmental degradation and declining food-crop yields are widely reported and officially recognized by the Kenyan government. Over much of the high and medium potential areas *per capita* land holding is already less than half a hectare. This is projected to fall to a quarter by the end of the century. In a number of districts there is already reported to be little additional land available for subsistence production.

An examination of the pattern of land holding reveals another side of the problem of the subsistence sector. In the areas of high potential, in particular, the pattern of land holding is far from even. Some 2,227 farms, 0·1 per cent of the total, owned 14 per cent of the cultivated land. Overall 32 per cent of the land is operated by 2·4 per cent of the farms. This represents the majority of the cash-crop sector and generally occupies the best land. Thus the subsistence sector has access to a little over half the land, much of it of lower potential. How did this situation arise?

During the colonial period Kenya was divided into the 'scheduled areas' and the 'native reserves'. In the former only Europeans were allowed to hold land, while the reserves into which the majority of the rural population were confined rapidly became overcrowded areas of subsistence production. By the late 1940s, 36,260 sq. km of land, mainly of the best quality, had been allocated to 3,500 European farmers and companies. In contrast 4 million Africans were restricted to 134,670 sq. km. This allocation removed the traditional sector's capacity for expansion and surplus production. During the 1950s, and more particularly the 1960s, after independence, some redistribution did take place in response to intense rural political pressure. The colonial legacy of Kenya remains, however. Large-scale commercial cultivation by individuals and companies still contrasts starkly with the overcrowded subsistence sectors.

(Based on material in Baker 1982 and Hunt 1984).

conditions. All too often the problems of Third World agriculture are attributed to agricultural practices or the environment, with little consideration of how the present situation arose. This is very much the case in the debate over the Sahel drought (figure 3.3). The proposed solutions to the problem are likely to reflect strongly the explanation adopted. This is discussed at greater length in chapter 4.

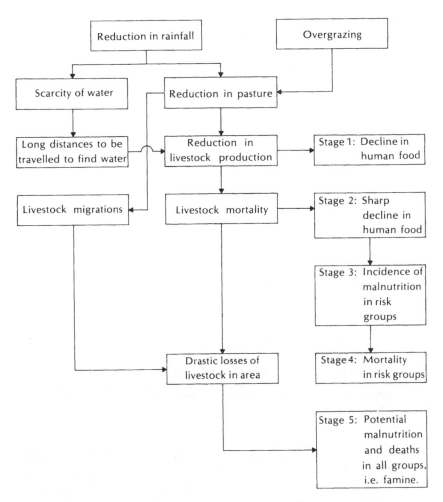

Figure 3.3 The process of breakdown: the Sahel nomadic economy during drought. *Source*: Fitzgerald (1978: 16)

In much of sub-Saharan Africa the traditional systems of shifting culti-vation (see chapter 2) have been unable to cope with the pressures on them. Evidence points to a decline in fallow periods almost everywhere. The situation depicted in figure 2.3 (c) is widespread, and in many cases the decline in yield and soil erosion are irreversible (plate D1). The increased demands made on the environment have not been accompanied by the development of more productive sustainable cultivation, and as Harrison

Table 3.4 Land holding in Java

1970	(International Labour Office study)
	41 per cent of rural households were either landless or had less than 0·1 ha of land.
1973	(Agricultural Census)

Hectares	*% of rural households*
Less than 0·5	57·4
0·5 to 1·0	24·0
Over 1·0	17·8
Average farm size: 0·64 ha	

1976	(National Social Economy Survey)

Hectares	*% of rural households*
Less than 0·05	38·2
0·05 to 0·5	37·5
Over 0·5	24·3

1977	(sample study of nineteen villages)
	53% of households were landless
1980	(Census)
	63·1% of households had less than 0·5 ha

Note: The trend towards smaller holdings and increased incidence of landlessness should be apparent from these figures. However, the problem of establishing trends from partial data, from different sources of unknown reliability, should be borne in mind.
Source: Hardjono (1983).

(1984) has noted, settled agriculture is being practised with the technology of shifting cultivation. However, it is not only in the ecologically fragile areas of the Third World that there are serious signs of established systems of cultivation being under stress. This is also true of many intensively cultivated Asian rice-growing areas which had previously shown a remarkable ability to increase surplus production and absorb population growth.

The worsening problem in rural Java is illustrated in table 3.4. Well over 50 per cent of the households are effectively landless, probably some 20 per cent of the population unemployed, and 35 per cent under employed. Access to land is declining with the development of large-scale commercial production. Traditional communal harvesting systems which entitled participants to a share of the crop are being rapidly replaced by commercial production using hired labour. Thus the landless or inadequately landed have lost an important element of the 'shared poverty' which characterized rural Java. Given the limited opportunities for rural employment, many people have no alternative but to migrate to the already overcrowded urban areas.

Key ideas

1 Third World agricultural systems have come increasingly under pressure as they have been incorporated into the world economy.
2 Incorporation disrupted or destroyed established mechanisms for coping with harvest failure, population growth, and the need for surplus production.
3 The establishment of production for the market brought Third World farmers a range of new uncertainties which they had no mechanisms for coping with.
4 The pressures placed on Third World agricultural systems have resulted in increased poverty and, in many cases, serious environmental damage.
5 The contemporary 'problems' of Third World agriculture can be fully understood only in the context of how they developed.

4
Rural poverty: explanations and solutions

The previous chapters have examined the nature, extent, and emergence of the rural sector. This chapter focuses on the various explanations that have been offered for the persistence of rural poverty and the equally varied rural development programmes that have been advocated. It should be stressed that rural development policies can be understood only in the context of the explanations of rural poverty that lie behind them.

The meaning of 'rural poverty'

Most studies of rural poverty define the poor by reference to single measures such as land holding or *per capita* annual income. These indicators are valuable in establishing 'poverty lines', quantifying the extent and depth of rural poverty, differences between areas, and change over time. However, even when used in combination they provide only a very partial view of rural poverty. It is possible to list the characteristics of the rural poor in an attempt to produce a more representative picture (table 4.1).

Poverty varies in depth, and while measures of land holding, food intake, or income enable us to draw lines between, say, the very poor and the not so poor, we have little understanding of what those differences mean to the people concerned. The poorer people are, the more central to their life is the struggle simply to survive. Life is uncertain and precarious.

Table 4.1 Characteristics of the rural poor

Landless	Low life expectancy
Too little land	Low income
Family too large	Irregular income
Malnutrition	Weak bargaining position
Ill-heath	Isolated, owing to poor communications
Uneducated	Preoccupied with survival
High infant mortality rate	Indebtedness

Tawney (1932: 71) wrote of the China of 1931 that 'There are districts in which the position of the rural population is that of a man standing permanently up to the neck in water, so that even a ripple is sufficient to drown him'. In addition the poor have little access to political power and have limited bargaining power with merchants, landlords, moneylenders,

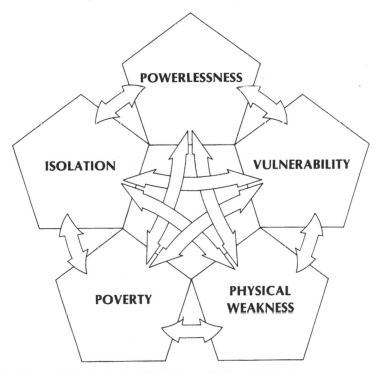

Figure 4.1 The reality of rural poverty. *Source*: Chambers (1983: 112)

Table 4.2 Explanations of rural poverty

Part of the underdevelopment of the Third World

Due to the extraction of surplus production under:
 (i) Colonialism
 (ii) Neo-colonialism
 (iii) International capitalism

Due to uneven exchange between:
 (i) Third World and Developed World
 (ii) Rural and urban sectors

Pre-modern social structures and attitudes

Ill-health and poor nutrition

Poor environments

Natural disasters

Shortage of resources

Distribution of resources

Wars and disruption

Rapid population growth

Over-population

Degradation of the environment

Inappropriate development policies

Bias in policies against the rural sector

Inefficient governments

Lack of education

---

or the state. The precariousness of existence and the preoccupation with survival makes political organization to increase their power difficult. Isolation, vulnerability, and powerlessness are all reinforced by the physical weakness that stems from ill-health and malnutrition. Chambers (1983) has shown how those factors are closely integrated (figure 4.1).

Explanations of rural poverty

Explanations for the persistence of rural poverty are extremely varied. The list in table 4.2 is by no means exhaustive – you might like to add to it. However, although all these have been advanced singly or in combination as explaining rural poverty, most people would agree that not all are equally important and indeed dismiss some entirely. Merely listing possible causes and then grouping, ranking, or eliminating them is, of course, a far from satisfactory approach. Three questions need to be asked. First, are some of these 'explanations' not 'causes' of rural poverty but 'consequences'? Second, are some 'underlying' causes which explain other problems? Third, do the causes of rural poverty lie in the rural sector itself or are they external to it?

Today most accepted explanations centre on either socio-economic factors or broadly environmental ones. These two groups encompass very different perpectives. The socio-economic approach, sometimes referred to as the 'political economy' view, sees the root cause of poverty in the distribution of wealth and power in society. In contrast, the environmental school focuses on shortage of resources, poor environmental conditions, and degradation of the environment. In practice many writers do not fall neatly into either of these two groups.

Earlier, and now generally discredited, views placed the causes of rural poverty firmly in the nature of Third World societies. In these explanations much emphasis was placed on the way in which pre-modern (non-Western) values and practices perpetuated 'backwardness'. Often these 'explanations' totally ignored the nature of rural systems of production and how they reacted to externally induced change (see chapters 2 and 3). Similarly, particularly during the colonial period, poverty was often depicted as self-inflicted. Individuals, families, and ultimately communities were poor because of idleness, drunkenness, gambling, unwise expenditure, incompetence, ignorance, and even lack of intelligence. Some of these explanations were overtly racist. Of course, at the level of immediate causes of poverty not all these factors can be completely dismissed. It is, however, generally agreed that they are valid only at a very superficial level. We cannot understand rural poverty without getting behind appearances.

Physical and ecological explanations

The relationship between poor environment and poverty has been long recognized, though few would now accept that this is a one-way causal relationship. Certainly some of the most impoverished rural communities are found in the poorest environments, for example in parts of the Sahel. Indeed, in much of sub-Saharan Africa a convincing argument for a large environmental contribution to poverty can be made. However, many extremely poor communities exist in environments that cannot be regarded as impoverished. Are the silt-rich deltas of the major Asian rivers, such as the Ganges and Brahmaputra, poor environments? Are the

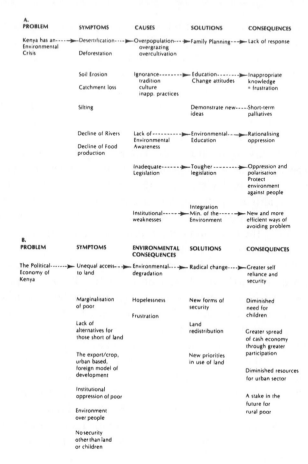

Figure 4.2 Two views of the situation in rural Kenya. *Source*: Baker (1981: 11, 21)

volcanic soils of Java? Yet West Bengal, Bangladesh, and rural Java have some of the world's deepest and most extensive poverty. Can we conclude on this evidence that poor environments can contribute to poverty but that elsewhere other explanations must be sought?

As discussed in chapter 3, rural communities can seriously damage the environment. For many writers unsuitable methods of cultivation combine with 'overpopulation' to create environmental degradation. This view is illustrated in figure 4.2 (a). In Kenya, as in many Third World countries, the official view of the causes of environmental damage is what has been caricatured as the 'rabbit syndrome'. That is, there are too many people trying to cultivate an already fragile environment. Thus poor environments and environmental degradation can, in some instances, be presented as self-inflicted.

Political economy explanations

Essentially the political economy view is that rural poverty is a result of processes which concentrate power and resources. These processes operate at all scales, from the international to the local (table 4.3 and figure 4.2 (b)). To understand the predicament of the rural poor they have to be seen in this broader context. Processes operating at the local scale then reinforce the national and international ones. A very simple example illustrates this. If the international price of maize falls, then the national wholesale price falls as a result, and merchants will try to pass the loss caused by the price fall down the chain. The smallest and least powerful merchants and farmers lose most. Furthermore, because many households need cash urgently to settle debts, purchase essential supplies, or meet an unexpected emergency such as medical costs, the crop may be sold 'green', before the harvest, at a substantial discount. The poorest households will be most vulnerable and will lose most in the process.

As discussed in chapter 3, the commercialization and incorporation of rural economics set in motion a process of differentiation. Existing

Table 4.3 The creation of inequality

International	Colonial exploitation Post-colonial unequal exchange Developed World investment and repatriation of profits	Division between rich and poor countries
National	Unequal exchange between rural and urban sectors Cheap urban food and low farm prices Main investment goes to the urban industrial sector	Political and economic power is concentrated in the hands of the urban middle class
Rural area	Local clites, large landowners, moneylenders, merchants and bureaucrats possess the power and most of the established resources. New resources tend to accrue to them as well	Polarization of rural society

resources become very unevenly distributed. New inputs (see chapters 5 and 6) tend to accrue to the wealthier and more powerful members of the community. The gap between the rich and poor widens. In extreme cases the extent and depth of poverty increase.

The political economy of the environment

It has increasingly been accepted that both the environmental and the structural views contain considerable elements of truth but are incomplete in isolation. As a result, attention has focused on the way structural processes interact with the environment. Thus, although the extreme environmental condition of the Sahel has resulted in increasing acceptance of environmental explanations for poverty, we need to know how such a situation emerged.

Extended periods of drought are not common in the Sahel, but the ecology of the region has become progressively more precarious over the last fifty years. This can immediately be attributed to over-grazing (see figure 3.3). Why has this happened? The straightforward answer would seem to be population growth and increased herd size. However, this is only part of the story. The growth of the agricultural population has also taken land out of grazing. In addition, rapid expansion of commercial agriculture has reduced the land available for grazing and in some cases for subsistence crop production. Indeed, large-scale cash-drop production, often involving multinational corporations, has taken land out of the local economy altogether. Pastoralism and sedentary cultivation have always been closely integrated. As they have been separated, principally owing to commercialization, each sector has become more vulnerable. As a result, a system that has been able to function in the Sahel environment is no longer able to.

the environment. Blaikie (1985) has suggested that the process of incorporation and commercialization extracts surplus production from the rural population. They in turn extract more from the environment (removing stored-up soil fertility and forests, for example), which may lead to soil erosion and environmental degradation. Figure 4.3 illustrates this process for Niger.

The nature of rural development

According to the World Bank (1975):

Rural development is a strategy designed to improve the economic and

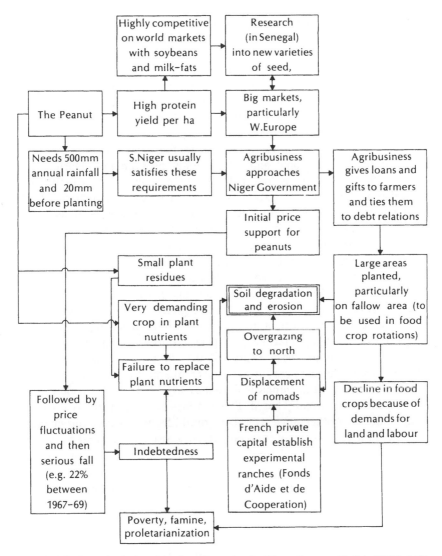

Figure 4.3 Causal relationships in soil erosion in Niger. *Source*: Blaikie (1985: 112)

social life of a group of people – the rural poor. It involves extending the
benefits of development to the poorest among those seeking a livelihood
in the rural areas. The group includes small-scale farmers, tenants and
the landless.

This definition reflects the way in which views of development have changed since the early 1970s. During the 1950s and 1960s development policies centred on 'growth maximization'. The poor were expected to gain from the 'trickle down' of the benefits resulting from overall rapid growth. By the late 1960s, however, it was realized that the benefits of rapid growth were not only taking too long to reach the poor, but would never reach most of them. As a result, the impact of rapid growth was a widened gap between the 'haves' and the 'have-nots'.

The need to reduce poverty and inequalities consequently became incorporated into national and international programmes. During the early 1970s such terms as 'redistribution with growth' or 'growth with justice' appeared. While this was a significant change it is also the case that most national and international policies have remained preoccupied with raising productivity. The 'redistribution' has been of new resources and incomes, not of old ones. In other words, the structures that lie behind the inequitable distribution of power and resource have been left intact.

However, it cannot be denied that the rural poor have begun to receive specific attention. In many instances this has involved pushing existing programmes into remote areas and poorer groups. Elsewhere there have been specific schemes aimed at the poor – these have included programmes for basic needs, for small farmers, marginal farmers or landless labourers. However, concern has been expressed that in practice many of these schemes benefit the less poor, or even the comparatively well-off.

As well as a change in emphasis since the early 1970s, rural development has become more broadly based. Programmes are less narrowly focused on agriculture or other single aspects of rural life. To a degree this change recognizes the complexity of rural poverty, discussed earlier, and has given rise to the term 'integrated rural development'. Again some critics have argued that, while there have been some notable successes, these programmes are often more integrated in name than in practice.

Rural development strategies

Rural development strategies may be grouped under three very broad headings (table 4.4). Most Third World countries have adopted policies that have a strong technocratic element. The central objective is to increase productivity, particularly in agriculture. This is the essence of the Green Revolution discussed in chapter 6. Such an approach leaves in place the structures that maintain inequality of incomes and access to resources. In consequence, particularly in the early stages, such policies tend to benefit

Table 4.4 Major types of rural development

Development strategy	Objectives	Major beneficiaries	Dominant form of tenure	Ideology	Representaive countries
Technocratic	Increase output	Landowning elite	Large private and corporate farms, plantations, latifundia, various tenancy systems	Capitalist	Philippines, Brazil, Ivory Coast
Reformist	Redistribute income (and wealth); increase output	Middle peasants, 'progressive' farmers	Family farms, co-operatives	Nationalist	Mexico, Egypt
Radical	Social change; redistribute political power, wealth, and output	Small peasants and landless labourers	Collectives, communes, state farms	Socialist	Vietnam, China, Cuba, Algeria

Source: Griffen (1974: 204).

Case study E

Mexican rural development: three contrasting views

1. The pro-peasant group (*campesinistas*) believe that rural poverty is a result of the way in which the peasant farmer has been treated by the government and the rural bosses (*caciques*). They advocate that more attention should be paid to the way in which traditional farming systems utilize scarce natural resources.

2. The left in Mexican politics attributes the poverty of the rural areas to international economic relations. Mexico's economic dependence on the USA has impoverished rural people in the interests of multinational corporations and agribusiness. The rural class structure is a reflection of these international development processes.

3. Less openly stated in Mexico is the view that there are too many people on the land. The bureaucracy and paternalistic politics have disguised the inefficiency of Mexican agriculture, while corrupt peasant leaders and government personnel have reduced agricultural productivity. This situation is contrasted with the most modern agricultural sectors, which are efficiently managed and make better use of the 'comparative advantage' conferred on Mexico by climate and proximity to the USA.

(*Source*: Redclift 1984: 80–1)

---

the richer farmers. This approach is sometimes justified on the grounds, first, of maximizing agricultural growth; and, second, that the redistribution of resources would be costly and politically unacceptable.

Radical strategies rest on fundamental social change which rejects capitalism and attempts to construct some form of socialism. This revolutionary approach is based on the acceptance of the structural causes of poverty. Thus the first stage in the elimination of poverty must be the removal of the processes that perpetuate it. In recent years much attention has focused on the rural development programmes of 'socialist' countries, most notably China and Cuba (see chapter 7).

The reformist approaches are essentially a compromise between the technocratic and the radical types of change. Such policies take a variety of forms but in essence they involve attempts to redistribute power, income, and access to resources. In practice the underlying structures are left largely intact. Chapter 5 examines attempts to redistribute existing land through land reform programmes and colonization schemes.

This chapter should give you a basis from which to consider the development presented in chapters 5 and 6. In practice the position of writers, planners, and government is often far from clear. All too often the debate over rural development appears to be taking place outside the Third World. Case study E summarizes the different views of rural development current in Mexico.

Key ideas

1 Rural poverty is extremely complex; little can be learnt about it merely by listing the symptoms. It is necessary to try to understand what poverty means to the rural poor.
2 A wide variety of 'explanations' have been offered for the persistence of rural poverty. Many of these 'causes' prove, on closer inspection, to be symptoms.
3 Very broadly, most contemporary explanations may be grouped under a socio-economic or environmental heading. Many writers now believe that investigation of the interaction between these two groups of factors is the most fruitful source of explanations.
4 There are many approaches to rural development. These may be classed under three broad headings: technocratic, reformists, and radical. They reflect very different views of the causes of rural poverty.

5
Redistribution of resources

Access to resources, particularly land, is a key factor in the differentiation of Third World rural communities. Land shortages are frequently presented as the result of exhaustion of the reserves of uncultivated land. While this is increasingly true in some parts of the world, in many instances the problem is not one of land shortage but of very inequitable distribution of ownership. Similar arguments of maldistribution rather than shortage can be advanced for other resources. This chapter discusses some attempts to redistribute new and existing resources through land colonization schemes, programmes of land reform, and the development of rural institutions.

Expansion of the cultivated area

Despite the efforts devoted to raising agricultural productivity (see chapter 6), expansion of the cultivated area remains, for most of the Third World, the most important means of increasing output. Even in India during the 1970s more than 20 per cent of the increase in cereal production resulted from expansion of the area planted. In many parts of South America and Africa more than 80 per cent of the increases of recent years has been due to expansion rather than intensification. This is particularly the case for basic food crops. As seen in figure 5.1, there are very considerable variations in the rate of increase of cultivated land. In regional terms the

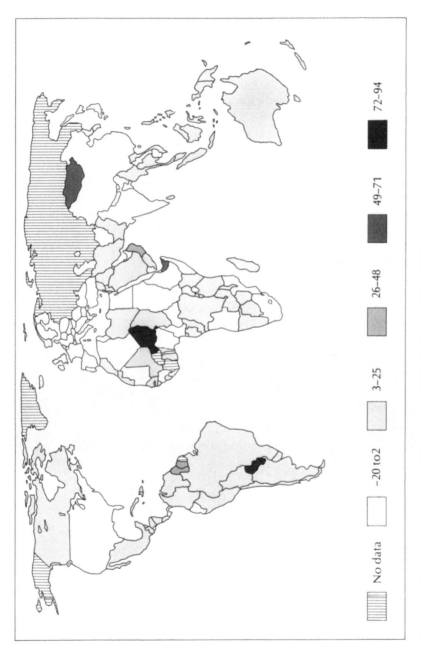

Figure 5.1 Percentage change in cultivated area, 1975–85. *Source*: FAO (1977, 1987)

No data

-20 to2

3–25

26–48

49–71

72–94

most spectacular increases are taking place in South America, largely through the clearance of the rain forest.

While land reclamation, often allied with irrigation and flood control, plays an important part in bringing new land into cultivation, the main increase comes from forest clearance. In Africa there is some 160 million ha of tropical rain forest; this is being cleared at a rate of 1·3 million ha a year. The even more dramatic clearance of the Amazonian rain forest is examined in case study F.

Some writers argue that land has to be cleared to increase Third World food supply. However, much of the newly cultivated land is unsuited to permanent cropping under prevailing methods. In many instances the increase in output brings little benefit to the majority of the rural population. Additionally deforestation can have serious economic and environmental consequences (see Avijit Gupta, *Ecology and Development in the Third World*, in this series).

The removal of forest cover can result in reduced local rainfall, increased run-off, flooding, and silting of irrigation works. Clearing of the tropical rain forest may result in long-term regional and even global climatic

Plate 5.1 Sawing planks for village house construction

Case study F

Colonization in Amazonia

The Amazonian rain forest is being cleared, primarily for agriculture, at a rate of 180,000 sq. km a year. At this rate the forest cover will be gone by

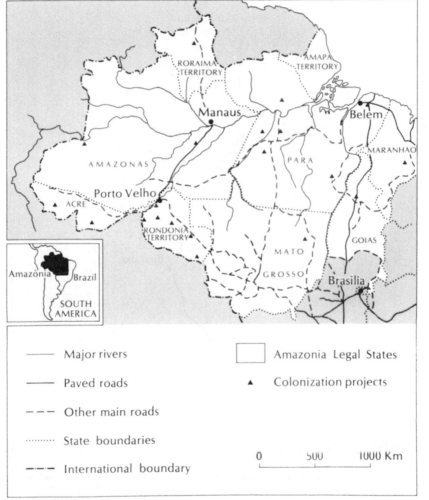

Major rivers		Amazonia Legal States

———— Major rivers

———— Paved roads   ▲ Colonization projects

– – – Other main roads

········ State boundaries

—·—·— International boundary

0   500   1000 Km

Figure F1 Colonization projects in Amazonia. *Source*: Leite and Furley (1985: 120)

Case study F *(continued)*

the end of the century. Settlement of the forest has been in progress for centuries, but since 1964 the Brazilian government has given offical encouragement to large-scale clearance. Colonization is planned and

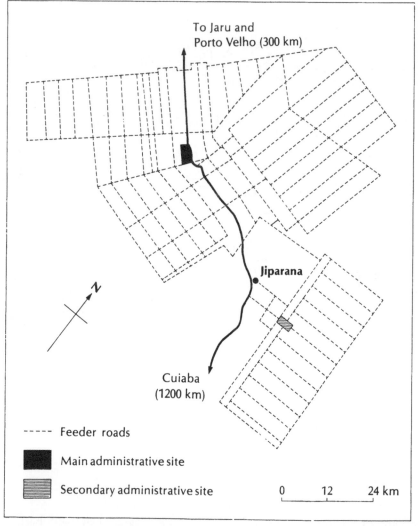

Figure F2 Plan of the Ouro Preto project. *Source*: Leite and Furley (1985: 120)

Case study F *(continued)*

controlled by INCRA (National Institute for Colonization and Agrarian Reform). This organization has juristriction over 350 million ha, over 12 million ha of which are undergoing active colonization. From the late 1960s large areas of rain forest were opened up by the construction of major roads, mostly notably the 5,000 km Transamazonian Highway (figure F1). These routes encouraged large-scale spontaneous clearance as well as providing a focus for official schemes. These were initially aimed at resettling families from the overcrowded north-east.

Under INCRA land is allocated to settlers in 100 ha lots (figure F2). Half the clearing should be left uncleared or planted to tree crops. In practice this rule is widely ignored. The 100 ha holdings are generally too large for colonists to manage without acquiring additional labour. As a result, a rapid polarization has taken place, with a few colonists establishing large holdings utilizing hired labour. The inequitable land-holding pattern of the long settled areas of Brazil are reproduced in the frontier areas.

In some areas as much as 90 per cent of the land has been sold to large concerns in lots as big as 1 million ha. Large-scale low-density cattle ranching and crop production are developed by these companies, many of which are multinational agribusinesses (see Chapter 3). These concerns

Table F1 The occupation of land in Santa Cruz, Bolivia, 1981

Farm size (ha)	No. of titles	%	Area (ha)	%
0–500	43,017	97·0	4,196,462	45·5
500–2,000	632	1·4	691,074	7·5
2,000–10,000	659	1·5	2,667,534	28·9
10,000+	70	0·1	1,676,735	18·1
	44,378	100·0	9,231,805	100·0

Notes: 1. These figures have been calculated from two main sources of data: land registration documents and land distribution under the 1953 Agrarian Reform Law. The data refer to titled land only, i.e. they exclude most of the land in the colonization frontier. It thus under-represents private occupation of land under 50 ha in area.
2. The figures refer to land titles rather than landowners. They fail to reflect multiple ownership of land titles by the same person or family. The concentration of land ownership for which titles exist is likely to be greater than the figures suggest.

Source: Redclift (1987: 124).

Case study F *(continued)*

---

often displace the smaller settlers, who remain as labourers or move farther into the forest. In general the concentration of land ownership increases the farther the area is from the clearance of virgin forest. Conflicts between ranching, family farms, and multinational agribusinesses are now of considerable intensity throughout the Amazon region. Such developments are not unique to the Brazilian colonization. Table F1 illustrates the distribution of holdings in the Santa Cruz 'colonies' of Bolivia.

In all the south Amazonian rain forest frontier areas serious problems of land degradation have emerged. In general the methods of cultivation are not sustainable under the prevailing environmental conditions. Soil fertility declines sharply, and attempts at large-scale mechanization are likely to engender soil erosion. Sustainable methods of cultivation have been developed, utilizing inter-planting, rotations, and diversified production, but these are geared to the needs of small family producers whose activities are incompatible with the conditions prevailing in the frontier areas. In addition the continued availability of uncleared land does not encourage farmers to think of long-term cropping. The uncertainty governing conditions in the frontier also militates against farmers making long-term investments in tree crops such as cocoa or coffee.

---

changes. For rural dwellers the loss of forest cover has serious and very immediate implications. The problems created by the shortage of fuel wood have been widely reported (see John Soussan, *Primary Resources and Energy in the Third World*, in this series). Less attention has been focused on the implications for rural handicrafts. A whole range of village activities are dependent on local timber, including house-buildings (plate 5.1), and the construction of equipment like ploughs, carts, looms, spinning wheels, and pestles and mortars (for hulling grain or grinding flour). The manufacture of the more specialized of these often forms an important additional source of livelihood for households. In addition, forested areas can provide a whole range of food and raw materials vital to the rural community in general, and the poor in particular. Nevertheless, increases in the cultivated area through large-scale forest clearance have failed to keep pace with rural population growth. In consequence, over large areas of the Third World *per capita* land holding has fallen.

Land colonization

Spontaneous colonization of land by groups and individuals (plate 5.2) is still in many countries the main source of expansion of the cultivated area. Until comparatively recently, government involvement in the process was generally limited to providing a framework of land law and administration for pioneer areas. In a few cases, most notably in the Dutch East Indies, migration to pioneer areas was officially encouraged. Increasingly however, colonization, land shortages, and attempts to preserve forest areas have resulted in much spontaneous clearance becoming illegal. Serious conflicts have resulted between government and the interests of forestry, large landowners, and settlers. The sensitivity of the issues involved has resulted in the problem being poorly documented. In many countries large areas are 'squatted'; the uncertainty over the right to cultivate the land permanently adds to the pioneers' difficulties and encourages unsustainable methods of cultivation.
encourages unsustainable methods of cultivation.

Table 5.1 illustrates the various motives that lie behind state-sponsored colonization. These frequently operate in combination and are in practice

Plate 5.2 Dry-season clearance of subtropical forest

Table 5.1 Aims of colonization schemes

Expansion of agricultural production

(a) For domestic consumption
   e.g. in Malaysia many of the federal land development schemes have been aimed at increased rice production

(b) For export crops
   e.g. expansion of coffee production in Peru and Brazil

To consolidate control over sparsely settled border areas
   e.g. late nineteenth-century settlement of the northern Japanese island of Hokkaido because of fear of Russian annexation
   e.g. many contemporary Peruvian, Brazilian, and Bolivian schemes have an element of this

To reduce overcrowding in long settled areas
   e.g. Indonesian 'transmigration' programmes
   e.g. Amazonian colonization aimed at reducing pressure on land in north-eastern Brazil

Diversion from demands for land reform
   e.g. Brazilian, Venezuelan, and Colombian programmes

Source: based on Dickenson *et al*. (1983: 117–18).

sometimes incompatible. There are a number of major Third World state-sponsored agricultural colonization programmes currently in operation. The largest schemes are those in Amazonia (case study F) and Indonesia (see David Drakakis-Smith, *The Third World City*, in this series).

Land reform

Land reform is a blanket term for a wide range of measures introduced in a variety of circumstances. Some reform programmes focus on single issues such as: the subdivision of large estates; the consolidation of small or fragmented holdings; the abolition or reform of tenancy; the removal of communal ownership and the establishment of private titles to land; and, in the case of radical change, the complete or partial removal of private ownership through land nationalization. Most programmes, however, involve a combination of measures.

While land reform, as its name suggests, is identified with the reformist approach to rural development outlined in the previous chapter, it is also a feature of technocratic and radical strategies. During the colonial period the establishment of private ownership of land was a key element in breaking down subsistence production. Subsequently the successful implementation of technocratic strategies has often rested on programmes of land consolidation. Similarly, land reform has been a basic demand of rural revolutionary movements and a major element of post-revolutionary development policy in, for example, China, Cuba, and Vietnam.

While few writers on, or practitioners of, rural development are against land reform, it is viewed in very different ways. Behind broad agreement on the need for reform there are often profound differences. For some advocates, land reform is almost a 'magic wand' which, by redistributing land in a more equitable manner, will reduce poverty and tension between the 'haves' and the 'have-nots'. Such reforms are sometimes presented as essential if there is not to be violent political change. The close links between poverty and the land-holding patterns presented in table 5.2 form the basis of these views. At the other extreme are those who see land reform as undermining the existing political and economic structures, reducing agricultural output, and threatening export earnings. Between the extremes are found those who, while not against reform, are concerned about how it can be carried out and what the results will be.

For many Third World countries, particularly in South America, the 1960s were the heyday of land reform. Since the mid-1970s many have questioned whether land reform can in isolation provide a solution to rural poverty. Where land reform is seen as having been successfully implemented – for example, in Peru, South Korea, and Taiwan – redistribution was initially accompanied either by stagnation or by an actual

Table 5.2 Small and large holdings in Third World countries

	Small holdings			Large holdings		
	Numbers (%)	Area (%)	Average size (ha)	Numbers (%)	Area (%)	Average size (ha)
Latin America	66·0	3·7	2·7	7·9	80·3	514
Africa	66·0	22·4	1·0	3·6	34·0	28
Near East	50·0	11·2	1·6	10·3	54·7	50
Far East	71·1	21·7	0·7	4·0	31·1	17

Definitions:

	Small	Large
Latin America	Below 10 ha	Above 100 ha
Africa, Far East	Below 2 ha	Above 10 ha
Near East	Below 5 ha	Above 20 ha

Source: Harrison (1981).

decline in agricultural output. It is increasingly accepted that reform must be accompanied by measures to raise the productivity of land (chapter 6) and ensure equitable distribution of inputs.

The Brandt report went further, suggesting that in countries such as Bangladesh land is in such short supply that redistribution would provide only very limited relief for the poor. Efforts should therefore be concentrated on raising productivity. However, land holding in Bangladesh is not only uneven but is becoming more so under the impact of population increase and the spread of new crop technology (see chapter 6). 'Distress sales' of land by the poor are balanced by land purchases by the middle and large-size farms which have been the main beneficiaries of the Green Revolution. In terms of alleviating poverty a case can be made for redistribution and controls to prevent the accumulation of large holdings. The benefits of such a programme would have to be balanced against the likely impact on agricultural productivity and output.

In evaluating land reform programmes, the conditions under which they took place must be carefully examined. The success of reform in Japan, South Korea, and Taiwan has resulted in similar programmes being advocated elsewhere, with little attention to the very different circumstances that prevail. This is illustrated in case study G.

Rural institutions and distribution

A wide range of programmes aimed at accelerating rural development and/or reducing poverty involve the development of rural institutions. These have been chiefly concerned with marketing, credit, the supply of agricultural inputs (such as seed and fertilizer), and extension advice.

Much attention has focused on the problem of rural credit. The high cost of traditional moneylending sources (rates of 20 per cent a month are common) locks households into a cycle of debt. Expansion of institutional credit through banks and co-operatives was expected to break this cycle, but in practice, while the expansion of such credit provision has been impressive, in many countries it has not reached those most in need. Institutional credit, often heavily subsidized, is normally available only to the creditworthy. The poor are usually excluded from such sources and remain trapped in a circle of debt and disadvantaged sale of produce (case study H and table 5.3). Thus poorer households are forced to borrow from informal sources to meet immediate needs. In contrast, wealthier families will obtain institutional funds for productive investment, and money for social activities, from moneylenders.

Case study G

Land reform in Asia: South Korea and the Philippines

The concentration of land ownership in Korea was extreme. In 1914 it was estimated that 1·8 per cent of rural households owned 51 per cent of the land. Evidence points to increased concentration during the Japanese colonial period (1910–45). A major land reform programme was initiated in 1949 and was largely complete by 1953. Under this programme holdings were limited to 3 ha, and landlords were obliged to turn their land over to tenants in return for compensation equal to one and a half times the value of the land's annual production.

Table G1 Distribution of cultivated land by farm size in Korea, 1974

Farm size (ha)	Farm households (%)	Cultivated area (%)
Landless	4·5	0·0
Less than 0·5	29·4	10·6
0·5–1	34·9	29·7
1–3	29·8	52·9
More than 3	1·4	6·8

Source: Douglas (1983).

The results of this programme are illustrated in table G1. It is generally agreed that this pattern has been maintained, with little reversion to tenancy and concentration of ownership, or increased landlessness. The long-term success of the South Korean programme has to be seen in the context in which it took place:

1 Disruption of the rural areas by the Korean War (1950–3). This eased the implementation of the reform.
2 Rapid inflation during the 1950s, which made the cost of compensation negligible and seriously weakened the position of the farmer land-owning class.
3 South Korea was not dependent on agricultural exports.
4 Large volumes of aid from the USA.
5 The ability of the urban industrial sector to absorb surplus rural population.
6 A wide-ranging agricultural development programme, particularly after 1972.

Case study G *(continued)*

---

7   Firm enforcement of the land reform regulations by a strong central government.

In the Philippines by the mid-1950s the concentration of land ownership was at least as extreme as in pre-war Korea. Some 42 per cent of the land was owned by only 0·36 per cent of the families. The series of land reform measures introduced since 1954 have had little impact. This has largely been due to:

1   The continued power of the landowning class.
2   The dependence of the Philippines on the large-scale production of export crops.
3   The understaffing and underfunding of the programmes.
4   Regulations not being enforced.

In 1972 President Marcos announced a wide-ranging programme. Before it was implemented the landowning interests were undermining it. Land was subdivided among relatives, tenants often being evicted in the process. In some cases tenants were forced to hide the fact that they were tenants. In addition, the proposals had serious shortcomings. First, the maximum holding size was 7 ha, thus reducing the scope for redistribution. Second, the land under export crops was excluded. This was nearly 40 per cent of the cultivated area. Third, the compensation payments were too high for the former tenants. Many fell into debt and had to give up land to the moneylenders, banks, or even the original owners. Finally, there were no agricultural development programmes from which the small farmers could benefit. As a result by 1986 only 13,590 tenant farmers had received title to 11,087 ha of land. In 1987 it was estimated that 80 per cent of the land was still controlled by 20 per cent of the families.

(*Source*: based on material in Douglas 1983 and Canlas *et al.* 1988)

The low price levels and weak bargaining position of farmers have given rise to a variety of 'market improvement schemes'. These range from various attempts at regulating weights and measures to the establishment of state marketing monopolies. Frequently such programmes have encountered serious opposition from merchants. In some cases co-

Table 5.3 Rural credit in Western Orissa, India: participation in the credit market by source of credit (% of borrowing household)

Farm size group	Source of credit				
	Formal only	Informal only	Both	Not borrowing	Total
Landless labourers	4·3	75·7	2·9	17·1	100·00
Marginal farmers	4·3	56·0	16·4	23·3	100·00
Small farmers	32·8	29·7	18·8	18·7	100·00
Medium farmers	42·5	10·0	27·5	20·0	100·00
Large farmers	60·0	13·3	10·0	16·7	100·00

Source: Sarap (1987).

operation among established traders has effectively curtailed marketing programmes. Efforts to give farmers 'market information' are cheap and attractive to development agencies. In isolation they have little impact. The widely reported response of merchants to complaints that the price quoted on the radio is higher is 'Then go and sell it to the radio station.'

While merchants are widely condemned as exploiting farmers and contributing to their poverty, they can be important elements in agricultural change. The diversification of north-eastern Thai agriculture (case study B) since the late 1950s has been very largely due to the promotion of new crops by merchants. Cassava spread rapidly during the 1970s as merchants quoted prices, gave advice on cultivation, and supplied cuttings. In many cases these developments took place in the face of opposition from the official agricultural extension service.

As is shown in the following chapter, extension services have played a vital role in raising agricultural productivity. In general, extension programmes have been least successful in reaching the poorest farms. They have tended to concentrate on commercial production rather than subsistence cultivation. Similarly most services focus on male farmers. In much of sub-Saharan Africa the basic food production is in the hands of women (see chapter 7). In the absence of wider changes all too often programmes involving the development of rural institutions tend to reinforce the existing pattern of inequality.

Case study H

The effect of seasonal variation in price levels

Farmers sell produce immediately after the harvest, at the time of the lowest price (figure H1) because of the need to repay debts. To survive during the rest of the year and plant the next crop it will be necessary to borrow again (figure H2). This circle of debt can become a spiral when unexpected expenditure such as medical bills, or a sharp fall in prices, increases the level of indebtedness.

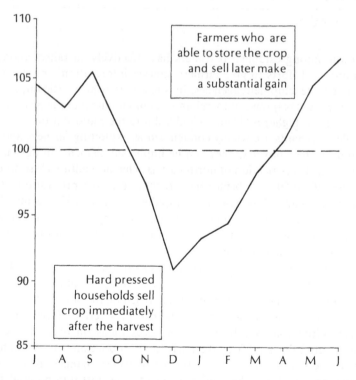

Figure H1 Index of seasonal variations in the price of rice in Bangladesh. *Source:* price data from Raikes (1981)

Case study H *(continued)*

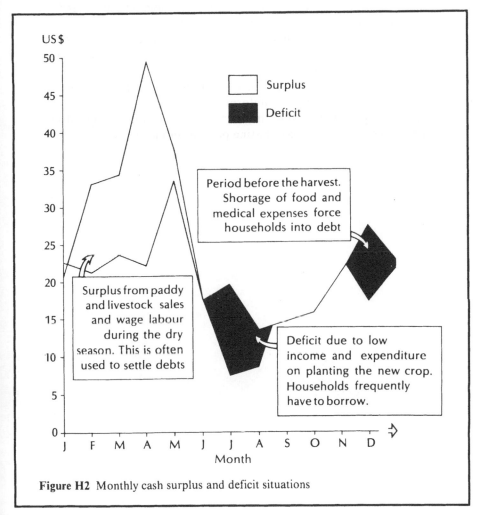

Figure H2 Monthly cash surplus and deficit situations

Key ideas

1 Expansion of the cultivated area remains, for most of the Third World, the main source of increased agricultural output. In many areas there are now serious land shortages, indicated by the falling *per capita* holdings and serious environmental damage.

2 In most countries land is very inequitably distributed; even when there is no overall shortage of land, many rural households are either landless or farm holdings that are too small to support them.

3 The poor are frequently unable to benefit from colonization schemes. Patterns of inequality prevalent in long settled areas are usually rapidly recreated in the frontier.

4 Land reform is a blanket term for a wide range of policies, induced with a variety of aims in far from uniform circumstances.

5 The development of rural institutions to supply credit and agricultural inputs frequently reinforces existing patterns of inequality.

6
Increasing productivity

The significance of yield increases in the expansion of Third World agricultural production is illustrated in table 6.1. Such increases have been most crucial in Asia and the Middle East and least important in South America and Africa. There is, however, considerable debate over the

Table 6.1 Contribution of changes in yield and area planted to increases in world cereal output, 1950–80 (%)

Developed	Developing	World
1950–60		
Area	−2	82	56
Yield	102	18	44
1960–70		
Area	−5	25	12
Yield	105	75	88
1970–80		
Area	42	16	25
Yield	58	84	75
1950–80		
Area	3	40	15
Yield	97	60	85

Source: Barr (1981: 1087–95).

magnitude of these changes. It may well be that the significance of rising yields has been exaggerated. The World Bank has estimated that since 1960 increases in yield have accounted for only 30 per cent of the expansion of production in South America and virtually none of that in Africa. This chapter focuses on the increases in the productivity of Asian agriculture (what is usually referred to as the 'Green Revolution') and examines the reasons for the limited impact of such developments in South America and Africa.

The Green Revolution

The term 'Green Revolution' is widely used but far from clearly understood. While most people connect the term with the introduction of high-yielding varieties (HYV) of certain food crops, particularly wheat and rice, the nature of the Green Revolution is far more complex. Three distinct meanings can be attached to the term:

1 A breakthrough in plant breeding that produced high-yielding food grains.
2 A package of technology, including fertilizer, insecticide, implements, water control, and HYV seed.
3 An agricultural development strategy in which the application of technology to Third World agriculture is central to the problem of increasing food production.

In many instances it is not readily apparent which or what combination of meanings is attached to the term. However, for most advocates of the Green Revolution approach it is the application of science and technology to Third World agriculture that is crucial. For the critics, however, the application of often inappropriate and mainly Western technology to Third World agriculture has created more problems than it has solved.

The origin and characteristics of HYV

The HYV around which so much euphoria erupted in the mid-1960s were the products of long periods of scientific plant breeding. The dwarf wheat introduced into India in 1965 was the result of work on Mexican agricultural experimentation stations since the 1940s. In later developments the International Maize and Wheat Improvement Centre (CIMMYT) was particularly important. Improved varieties of wheat were a key element in the rapid expansion of Mexican agricultural production. Between 1942 and 1964 output expanded by over 5 per cent a year and the country moved from dependence on imports to become a net exporter. It is significant that the increase rested almost entirely on the efforts of a small number of large-scale irrigated farms. The small-scale Mexican producers (*ejido*) were almost entirely excluded from these developments.

The Mexican varieties that had been developed by the early 1960s were characterized by a high response to fertilizer, a short growing season, non-photo-period sensitivity (they could be grown in any part of the world and at any time of the year where temperature and moisture were adequate), a resistance to disease (particularly 'rust'), and their short stiff stems, which reduced the incidence of the plant collapsing under the weight of wheat ears.

The dwarf varieties exported to India in 1965 produced yields of 4,450 kg/ha under experimental farm conditions. In contrast, under similar conditions the improved indigenous Indian varieties produced only 3,200 kg/ha. Given that in the early 1960s Indian wheat yields averaged 1,200 kg/ha, the appeal of HYV to planners is obvious. Initially the dwarf varieties were found to be remarkably suited to the conditions prevalent on the larger farms in the most advanced agricultural areas of northern India (case study I).

The apparent success of the wheat developments stimulated international interest in similar developments for rice, the grain on which the largest proportion of Third World people are dependent. However, unlike wheat, much indigenous rice either responded to fertilizer poorly or, in some cases, negatively. In 1960 the International Rice Research Institute (IRRI) was set up at Los Banos in the Philippines. While CIMMYT consolidated principally Mexican developments, IRRI drew on the very successful long-term breeding programmes of a range of countries, notably China, Japan, and Taiwan. The first release of HYV rice, IR 8 (International Rice 8), took place in the Philippines in 1966. Yields of up to 6,400 kg/ha were obtained under ideal conditions at a time when the Philippine average was only 1,300 hg/ha. The introduction of the new varieties took place under the auspices of the Rice and Corn Production Consultation Council, which co-ordinated the activities of the separate agencies responsible for irrigation, marketing, extension, research, credit, co-operatives, land reform, and soil surveys. Implementation, which involved training extension workers and farmers, was concentrated not only in the most favourable areas but also on the most receptive farmers.

The Philippine programme was seen as successful and many observers

Case study I

The first phase of the Green Revolution: wheat-growing in Ludhiana

The Punjab is the area most closely identified with the gains of the Green Revolution. During the 1960s and early 1970s agricultural change, in many parts of the state, was both rapid and spectacular (table I1). In addition, several thousand tube wells and large numbers of small machines, such as seed and fertilizer drills and threshing machines, appeared. These developments were portrayed in the early 1970s as adding up to an agricultural transformation. Areas like Ludhiana were seen as the spearhead of an Indian agricultural revolution and a model for Third World agricultural development in general.

Table I1 Agricultural change in Ludhiana

	1960-1	1968-9
Percentage of the land irrigated	45	70
Fertilizer application (kg/ha)	19·7	271·3
Area under HYV (ha)	68·8	169,971·7*
Yields (kg/ha)	1,552·3	3,676·3
Tractors	n.a.	2,500–5,000

*Ninety per cent of the wheat area.
Source: Frankel (1971).

In the early 1960s, however, Ludhiana was already one of the most advanced agricultural areas in India. Even by the standards of the generally highly commercialized Punjab wheat-growing areas it was outstanding. Farms were generally large, the area having experienced widespread land reform during the 1950s (table I2). Thus 80 per cent of the holdings were over 4 ha, compared to 42 per cent of the Punjab as a whole. While some 46 per cent of farmers rented some land, only 4 per cent were pure tenants. The percentage of the land irrigated was also one of the highest in the country.

---

considered it a model for Third World agricultural development. By the mid-1970s average rice yields had risen by 38 per cent to over 1,800 kg/ha. This figure concealed very considerable variation between areas and farms. The most advanced irrigated areas had by far the highest yields (figure 6.1). The early HYV of rice shared the general characteristics of HYV of wheat

Case study I *(continued)*

Table 12 Farm size in Ludhiana

Area (ha)	% of farms	% of cultivated area
Less than 2	4	1
2–4	16	9
4–8	43	35
Over 8	37	55

Source: Frankel (1971).

In other respects, also, the area was far from typical of rural India. The literacy rate was 36·3 per cent, compared to 28 per cent for India as a whole. Rural industry employed 35 per cent of the labour force – one of the highest incidences in the country. Indeed, the absorption of labour into industry was such that the 1961 census reported a general shortage of agricultural labour in Ludhiana.

Even taking the exceptional nature of the Ludhiana area into account, the changes that took place during the 1960s were remarkable. However, the limited number of other areas in India or the Third World as a whole experiencing similar developments led observers to question the general applicability of the model. Moreover, it is important to realize that the gains of the Green Revolution in Ludhiana were by no means evenly distributed. By the early 1970s studies suggested that while the 80 per cent of farms over 4 ha were generally better-off, those of less than 4 ha were often worse-off. As HYV spread during the late 1960s, the average yield fell from 4,400 kg/ha in 1966–7 to 3,360 in 1968–9, a reflection of the increasing number of small farms cultivating the new varieties. Ludhiana, the 'showpiece' of the Green Revolution during the 1960s and 1970s, revealed the limited applicability of the new technology.

(Based on Frankel 1971)

---

except that they were much less resistant to disease, particularly 'blast', so called because affected fields look as if they have been scorched. IR 8 and its immediate successors were very sensitive to environmental conditions and demanded a very high level of water control, land preparation, and tending. Adaptation to farm conditions was thus much more problematic

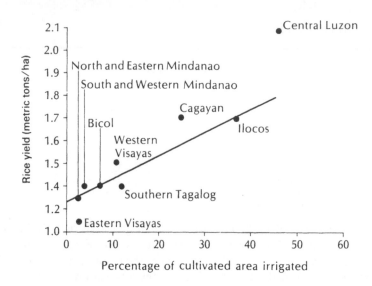

Figure 6.1 Rice yield and irrigation in the Philippines. *Source:* data from ILO (1974)

than for wheat, and the areas in which they could be effectively grown correspondingly fewer.

The early HYV of wheat were criticized because of their poor eating, milling, and nutritional qualities. While some authorities believed too much has been made of these faults, they were certainly far more significant for rice. There is evidence of an inverse relationship between protein content and yield of some HYV of rice. In parts of Bangladesh it has been suggested that malnutrition has increased as HYV have spread, particularly among children. The adoption of HYV of wheat and rice commits the farmer to annual purchases of seed. Re-sowing from harvested grain, particularly of rice, results in a rapid loss of HYV qualities. This limits the possibilities of participation in the Green Revolution by subsistence and semi-subsistence farmers.

Since the early 1970s the main thrust of rice breeding has been the production of varieties that are more suitable to farm conditions (plate 6.1). This recognizes the much greater variety of field conditions, cultivation systems, and cropping preferences found in rice production, as well as the generally lower level of development of most rice-growing areas. (Case study J illustrates the variety of rice types developed in Malaysia.)

Plate 6.1 IR 36, now the most widely grown of the IRRI-produced varieties, is lower-yielding than IR 8 but more resistant to disease and pests

The Green Revolution strategy

The mid-1960s was a period of crisis for Indian agriculture. Long-term growth in production appeared to be faltering, as the possibility of extending the area cultivated declined and intensification was making only

Case study J

The development of HYV rice in Malaysia

Malaysia's rice-breeding programme sought to combine the preferred eating qualities of traditional 'indica' varieties, grown in the Asian tropical and sub-tropical regions, with the high-yielding, early-maturing, low photo-period-sensitive, short-stiff straw, and fertilizer-responsive 'japonica' varieties (with a sticky eating quality) which had originated in the temperate zones of China. The first fifteen years of Malaysia's paddy-breeding research resulted in the development and release of two varieties – Malinja, in 1964, and Mahsuri, in 1965 – that were high-yielding and had improved milling and eating qualities.

During the next ten years, eight additional high-yielding varieties of paddy were developed and officially released through Malaysia's Department of Agriculture and Agricultural Research and Development Institute, each with slightly differing quantities from each other, related to disease resistance, taste, yield, etc. Despite these developments, officially released varieties remained upopular with the majority of farmers. Initially HYV spread rapidly from an estimated 10 per cent of the planted area in 1966 to 36 per cent in 1971–2; subsequently expansion has been very slow and farmer-selected varieties still account for more than two-thirds of types planted.

The early releases of HYV in Malaysia had serious shortcomings; they

---

limited progress. Indian grain imports had risen steadily since 1960, the country becoming increasingly dependent on food imports and, in particular, American aid. Disastrous harvests in 1965–6 and 1966–7 brought matters to a crisis. Grain imports reached 10·3 million metric tons, over 16 per cent of production. Under these circumstances it is easy to understand the enthusiasm of Indian planners for a new HYV-centred agricultural strategy, and the general euphoria which resulted in such slogans as 'an agricultural revolution', 'a seed–fertilizer revolution' and, most lastingly, 'a green revolution'.

Although the suddenness of the change in direction of Indian agricultural strategy can be exaggerated, it was very real. Since independence the main thrust of agricultural development had been the more intensive use of existing inputs. Quite simply this involved the more effective use of land and labour, coupled with an increase in the area

Case study J *(continued)*

were prone to disease, had poor eating and milling qualities, and their short stems made cultivation difficult. Why, despite the gradual elimination of these problems with each new release, do the majority of farmers prefer varieties that they select themselves? In part the answer lies in the unsuitability of HYV for the non-irrigated areas. These, however, account for less than 25 per cent of the rice area. More important is the difference between official and farmer seed selection.

Generally the official selections were based on one or two characteristics, initially yield and photo-period sensitivity. Subsequent selections attempted to make the varieties more suitable to field conditions. In contrast, farmers tended to use a much wider range of characteristics: for example, resistance to flood, drought, and disease; the quality of grain (this might be very different for sale and for domestic consumption); ease of harvesting and threshing; the length of the growing season; and yield.

The results of the farmers' selections were generally rice types that satisfied all their criteria and were well adapted to very local conditions. The official releases, in contrast, were adapted to rather more 'average' conditions. More recent developments of official varieties have moved towards selection based on local conditions, thus recognizing the importance of the farmers' approach.

(*Source*: Taylor 1981: 73–6)

cultivated. There had also been considerable investment in irrigation as well as some development of fertilizer application and improved crop varieties. The new strategy was to concentrate on the raising of productivity through the application of new inputs. Initially this new technology – HYV seeds, fertilizers, pesticides, water pumps, and tube wells – was imported.

The early post-independence strategy was extensive, and resources were spread as widely as possible so as to achieve an even distribution of benefits. Concern among Indian planners over the limited intensification that was taking place, reinforced by advice from the Ford Foundation, resulted in a winding down of the extensive programmes and the initiation, in 1960–1, of the Intensive Agricultural Development Programme (IADP).

This represented a fundamental change in the underlying philosophy of Indian rural development. The principle of an equal spread of limited

Plate 6.2 Water pumps, such as that shown above, have replaced more traditional methods of obtaining scarce water supplies in many parts of the rural Third World

resources was replaced by their concentration in the already most favoured areas. The IADP 'package' included improved seeds, fertilizers, pesticides, and advice on cultivation. Under the fourth Five Year Plan (1969–74) this programme became known as the NAS (New Agricultural Strategy). Essentially the development effort became concentrated on only 10 per cent of the cultivated area and principally on wheat.

HYV and farm size

Much debate has centred on what some authorities have seen as the failure of small farmers to take advantage of the Green Revolution. Such farmers are frequently described as 'resistant to agricultural change' or 'prone to adopt HYV packages in a partial and unsatisfactory manner'. Others have argued that it is not the farmers that are at fault but the new technology itself. This has given rise to an equally broad debate over whether the HYV packages are 'scale-specific', i.e. more suited to some sizes of farmer than others.

Large farms generally have greater access to effective irrigation and the capital necessary for land improvement and water control. With respect to loans, larger cultivators have simply been recognized as more credit-worthy. In some instances the size of institutional loans has even been tied to size of holding. In the Punjab, for example, the extension of irrigation by the development of tube wells is scarcely possible for small farms. The reduction of the land-holding qualification for tube-well loans during the late 1960s and early 1970s (from six hectares to two) did little to increase the take-up rate for farms below six hectares. For much of the state, farms below eight to ten hectares are considered 'sub-optimal' for tube wells. In the early 1970s this effectively excluded 20 per cent of farms developing more effective irrigation.

Similarly, the development of irrigated HYV wheat cultivation in north-west Mexico was accompanied by a very rapid increase in the average size of holdings as larger landowners either bought out the smaller, or rented their land. In the coastal area of Hermosillo in 1956 the average size of farm was 107 ha. With the rapid expansion of government-financed irrigation into the desert the average size of holding grew to 320 ha by 1971. In the areas of north-west India and north-east Pakistan land prices rose sharply in villages that adopted HYVs and the average size of holdings generally increased. While the concentration of land ownership remains slight compared to the Mexican case, most studies confirm that in areas where HYV have been established the average size of farms and the number of landless labourers have increased. However, this process of 'polarization' has been much slower than many observers expected. It must be remembered too that the Indian rural population was far from homogeneous prior to the Green Revolution. Differentiation of the agricultural population in terms of access to land was already marked in many areas (see table 3.3). The Green Revolution generally accelerated the process. The introduction of the Green Revolution technology has often seriously disrupted rural society. Migration to urban areas has increased and violence has flared in many areas, notably in parts of Java, the Punjab, and Nigeria.

Irrigation and the limits of the 'new technology'

The HYV of wheat and rice must be viewed as an integral part of a new technology package. It is generally agreed that the level of yields for HYV varies according to the amount of fertilizers used, the adequacy and availability of soil moisture (this usually implies irrigation and flood

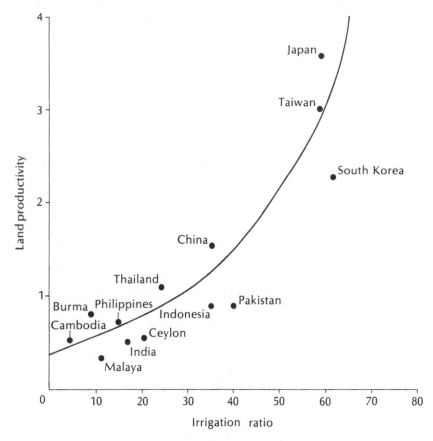

Irrigation ratio = $\dfrac{\text{irrigated arable land and land under permanent crops}}{\text{arable land and land under permanent crops}}$

Land productivity = $\dfrac{\text{total cereal production}}{\text{arable land and land under permanent crops}}$

Figure 6.2 Land productivity and irrigation ratio. *Source*: Bray (1986: 63)

control), and the quality of land preparation and crop tending. In wet rice cultivation the degree of water control has considerable influence on the proportion of fertilizer that reaches the plants. Poor control results in increased loss through leaching, run-off and, when the fields dry out, to the atmosphere. As noted above, HYV have generally been developed to maximize responsiveness to fertilizers under ideal water conditions; consequently a wide range of studies of wet rice cultivation has revealed the substantial increases in yields that results from irrigation even in the absence of fertilizer application. Figure 6.2 shows the relationship between irrigation and land productivity that prevailed in Asia before the Green Revolution.

The development of effective water control improves the growing environment for rice, reduces crop loss and damage, and stabilizes the growing period to that most suitable (note the problems of rain-fed rice cultivation discussed in chapter 3). In addition, the improved control over environmental conditions encourages the farmer to intensify. In the

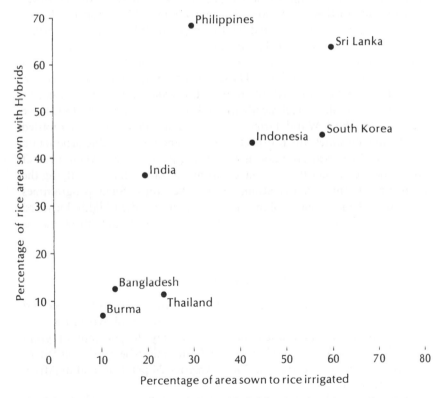

Figure 6.3 High-yield varieties of rice and irrigation in selected countries. *Source*: Grigg (1984: 169)

absence of fertilizer and HYV the establishment of effective water control has increased yield by as much as 50 per cent. Furthermore, the stabilization of the main season crop is a necessary first stage in the development of effective multiple cropping.

Water control may, therefore, be considered the most important element in the establishment of HYV varieties of wheat and rice in most of the Third World (figure 6.3). The new technology thus spread rapidly in areas where irrigation and commercial production were either well established or rapidly developing. Outside these comparatively restricted areas, progress has been limited. Even in Malaysia, where over 80 per cent of the paddy is irrigated, HYV have been adopted only in the most developed areas (see case study J). Much of the Third World experiences serious annual or seasonal water shortages (figure 6.4), and only limited progress has been made in the development of irrigation (figure 6.5), with restricted potential for further expansion . Currently some 112 million ha, 14 per cent, of Third World cultivated land is irrigated. While some authorities consider that it is *technically* feasible to irrigate another 400–500 million ha it may well not be *economic* to do so. In general the best irrigation sites have already been developed. Those remaining are likely to be costly, difficult sites where the risk of complications and even failure would be greater. Even if all the technically irrigable land were to be developed, the majority of Third World farmers would remain dependent on rain-fed cultivation, on which any prospects of higher and more reliable production levels rest. Increased attention is thus being paid to the development of crops and techniques that 'fit' the environment rather than modifying the environment, through irrigation, to suit the crops. Such programmes necessitate detailed and prolonged investigation of often highly localized environmental conditions and agricultural practices. To date progress has been limited.

Food production and the limits of the new technology

The crops that have formed the mainstay of the Green Revolution, rice and wheat, are simply not grown by large numbers of Third World farmers. Even in India rice and wheat account for only 70 per cent of grain production. For Asia as a whole, in 1985, rice and wheat represented 77 per cent of grain production, in South America 24 per cent, and in Africa only 10 per cent.

The food of the Indian poor consists of the so called 'coarse grains', such as millets and sorghum, or chickpeas and lentils. They are of major

Figure 6.4 Water surplus and deficiency. *Source:* Barrow (1987: 29)

(mm/yr)

> 1,000
< 1,000 ⎫ Surplus
> 0 ⎭

< −1,000
< 0 ⎫ Deficiency
> −1,000 ⎭

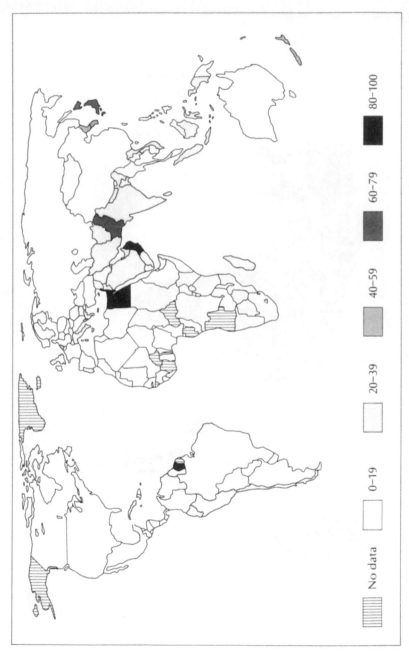

Figure 6.5 Irrigated area as a proportion of permanently cultivated land. *Source:* Barrow (1987: 29)

No data   0–19   20–39   40–59   60–79   80–100

1965   1985

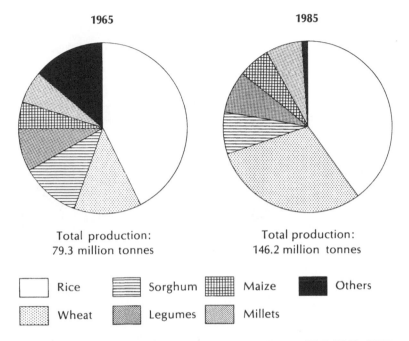

Total production: Total production:
79.3 million tonnes   146.2 million tonnes

☐ Rice   ☰ Sorghum   ▦ Maize   ■ Others

▨ Wheat   ▨ Legumes   ▨ Millets

Figure 6.6 Indian grain production, 1965 and 1985. *Source:* FAO (1967, 1987)

importance in the drought-prone, unirrigated areas of central and north-western India. Limited interest has been shown and even less progress made in producing improved varieties of these crops. Thus since the early 1960s the contribution of coarse grains to Indian food production has fallen by 10 per cent (figure 6.6). This is a reflection of static, and in some cases declining, yields as well as displacement by wheat as irrigation has spread. However, the poor often cannot afford wheat, and in any case the nutritional value of coarse grains is greater.

Similarly, in much of Africa food production is dominated by coarse grains (figure 6.7). The cultivation of rain-fed sorghum and millet accounts for an estimated 80 per cent of the planted area in the Sahel belt. Research into these crops has begun only recently and has had little impact. Some short-growing-season varieties of sorghum have been produced; this highly drought-resistant crop can take up to nine months to mature. These have not been free of problems, as Timberlake (1988) has observed:

In Ethiopia, peasants were offered a sorghum which matured in three

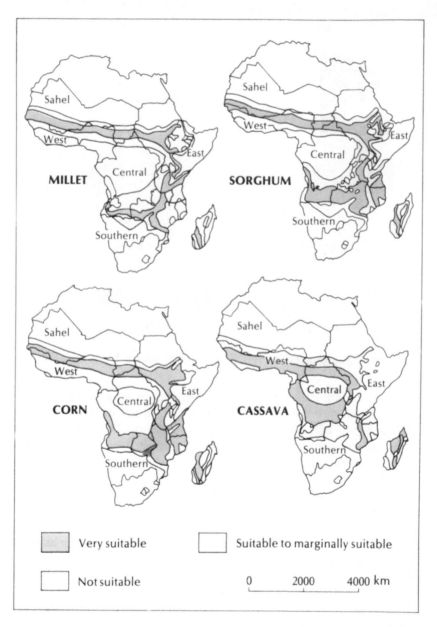

Figure 6.7 Suitable areas for rain-fed production of millet, sorghum, maize, and cassava in Africa. *Source*: Heibsch and O'Hair (1986: 178)

months. They would not plant it. In the highlands farmers use sorghum stalks as a roofing material and both the leaves and stalks as fodder for livestock. The fast-growing sorghum did not have the right sort of stalks and leaves.'

The Ethiopian example is by no means an isolated one. Many crops have vital secondary functions in the farm economy. To a partially informed plant breeder rice is solely a grain crop but to the farmer it is also a source of straw for animal fooder, husk for making charcoal, a 'mulch' to reduce evaporation in vegetable crops, or a building material when mixed with mud.

For much of South America and Africa the staple food crop is maize. Substantial progress has been made in maize varieties, particularly in Mexico. However, the main impact has been on large-scale irrigated and/or export-orientated production. Less than 25 per cent of the South American maize area is planted to improved varieties. In Africa, although hybrid maize has been introduced into parts of western Kenya, Zambia, and Ethiopia, the vast majority of subsistence and semi-subsistence farmers still grow traditional varieties. As with wheat and rice, hybrid maize has not proved universally acceptable to farmers. In the limited number of cases where suitable improved crop types have been introduced – for example, drought-resistant maize in western Kenya and green beans in Uganda and Tanzania – farmers have adopted them rapidly. However, in sub-Saharan Africa much work has gone into attempts to introduce HYV of wheat and rice. This ignores the realities of African food production. In addition the attempts to introduce Asian rice types into the established rice-growing areas of West Africa have largely ignored the indigenous types that are well adjusted to the local environmental conditions.

The discussion in this chapter has focused on grain production, but this is not the basic foodstuff of larger numbers of Third World people. This is particularly true in Africa, the Pacific, South America, and the Carribean, where root crops (cassava, sweet potatoes, potatoes, yams, taro) and tree crops (sago, plantains, breadfruit) provide the basic food. Even in Asia root crops are by no means an insignificant source of food. In China root crops, principally sweet potatoes or potatoes, represented 29 per cent of basic food production in 1985. Similarly cassava and sweet potatoes accounted for 23 per cent of basic food production in Vietnam and 22 per cent in the Philippines. Despite this importance, non-grain crops have generally received even less attention than coarse grains. Sometimes they

Plate 6.3 Cassava. Some 300 million people depend on this crop for their basic food

do not even appear in agricultural statistics. In Zambia the crop remains unrecorded despite being cultivated by over half the rural households.

Cassava (sometimes called tapioca or manioc) is by far the most important non-grain source of staple food in the Third World (plate 6.3). It is estimated that it forms the basic food of 300 million people. The crop will grown in a wide range of conditions, is resistant to drought, and produces good yields on infertile, acidic soils. Additionally cassava produces more calories per hectare than any grain crop. Given these characteristics, the spread of the crop as a cheap source of food in many parts of the Third World is scarcely surprising. In recent years the commercial production of cassava has increased, primarily for animal fodder, although the crop does have a number of industrial uses. The main expansion has been in South East Asia for fodder markets in the EEC, Japan, and South Korea. This development has stimulated interest in improved varieties and cropping patterns. However, little of the work has benefited the food-producing sector.

Case study K

The Variegated Grasshopper problem in southern Nigeria

The Variegated Grasshopper (*Zonocerus variegatus*) is found throughout the forest zone and wetter savannas of west and west-central Africa. The adult insect will attack many types of vegetation, including cassava, vegetables, and some tree crops. Cassava is the crop most at risk, because it takes from twelve to eighteen months to mature and thus remains in the ground right through the dry season. Crops such as maize and beans planted early on moisture-retentive soils are also vulnerable. Damage to these crops is especially significant since they are often planted to relieve food shortages during the 'hungry season'.

A study of the Zonocerus problem was undertaken in southern Nigeria in the early 1970s, after a decade or so with several major outbreaks. Some parallel studies were undertaken of local knowledge concerning the insect and its habits. Initially there was some suggestion that the Zonocerus problem might require highly organized control work along the lines pursued in the case of the Desert Locust.

Aspects of Zonocerus ecology familiar to farmers in southern Nigeria include knowledge of the life cycle of the insect, egg-laying behaviour and typical egg-laying sites, and factors influencing mortality rates. Many farmers hypothesize a general correlation between rainfall irregularities and fluctuations in grasshopper numbers. In a few cases, farmers interviewed had already anticipated the main pest control recommendations of the Zonocerus research project, namely, to mark out and dig up egg-laying sites, although these initiatives had not yet proved very successful, because they had only been undertaken on an individual, not a community, basis.

The particular significance of the Zonocerus case is that it illustrates well the potential advantages of a research partnership between scientists and farmers. Some discoveries made by the research team were beyond the scope of farmers, because they required laboratory facilities. Other pieces in the jigsaw were already in the possession of farmers before the formal research project began. Finally, farmers had some items of useful information for which they were the main or only possible source: for example, the relative significance of damage to minor, but locally significant, crops (e.g. fluted pumpkin in eastern Nigeria), and oral historical information about the timing and severity of previous plagues of Zonocerus.

Case study K *(continued)*

> The recommended strategy for controlling Zonocerus is that egg-laying sites should be identified and marked while in use, and later dug over, thus destroying egg cases by exposure to sunlight. Few farmers would have more than two or three such sites to deal with, and the work might take a few hours. The key to a control strategy of this sort is to ensure co-ordination of effort over a sufficient area. Treating one farm but not its neighbour would have little effect. Clearing the egg-laying sites from a block of farms reduces grasshopper numbers by 70–80 per cent in the following year. It is clear from this case study that many farmers already have much of the knowledge needed to undertake such a control programme, with addition, perhaps, of some external co-ordination by the extension services.
>
> (*Source:* Richards 1985: 146–9)

The lack of interest in cassava has been reinforced by the crop's reputation for exhausting soils. In some cases colonial and post-independence agricultural departments attempted to discourage cassava cultivation. However, as is frequently the case, it is the methods of cultivation and the pressures on poor and limited land that are the problem rather than the crop itself. More interest has been shown in the development of improved varieties of potatoes. In the Andean states of South America, particularly Equador, Colombia, Peru, and Bolivia, potatoes are a major food. In recent years improved varieties and cropping methods have been developed.

Key ideas

1 In the Indian sub-continent increased production of wheat and rice has substantially reduced dependence on food imports.
2 In the areas where the HYV have been successfully introduced the small farmers and landless labourers have often received little benefit. Additionally the displacement of coarse grains by wheat has reduced the supply of the basic food of the poor.
3 The spread of the new technology has been limited to areas of the Third World where there is sufficient moisture (usually necessitating irrigation), and the crops are suitable. Large areas, particularly in

Africa and Latin America, have been almost entirely excluded from the benefits of the Green Revolution.

4 For much of the Third World the need is for the development of rain-fed crops and cropping techniques.

7
Assessments, achievements, and alternative approaches

How successful have rural development programmes been? This is a far from easy question to answer. The evaluation procedures of both national and international development agencies have been widely criticized. All too often reviews are conducted by those who have a vested interest in showing that programmes are successful. One alternative to trying to evaluate particular policies and programmes is to examine the trends in Third World rural poverty. What impact have thirty years of rural development planning had?

Trends in Third World rural poverty are far from clear. In addition to the measurement problems discussed in chapters 1 and 4, there is the difficulty of obtaining comparable data over time. Trends are by no means uniform, and very different conclusions can be drawn depending on where, and at what scale, the study was undertaken. Indeed, it is possible to find evidence to support a wide range of assessments even for the same countries. In examining this whole question one must ask: who conducted or commissioned the study? Why did they do it? What interest did they have in the results?

Even if the more optimistic views on the reduction in the *proportions* of Third World people living in poverty are accepted, the *number* of poor has not diminished (table 7.1). While some national studies indicate a decline in the number of people below the poverty line, the degree of polarization may well have increased (table 7.2.).

Table 7.1 Population lacking basic needs (millions)

Region	1974 Total No.	1974 % of total	1982 Total No.	1982 % of total
Latin America	94	30·6	86	23·2
Near East	40	26·0	36	18·0
Asia	759	53·0	788	60·0
Tropical Africa	205	67·6	210	54·0
All developing countries	1,098	56·0	1,120	47·0

Source: Hopkins (1980).

Table 7.2 Thailand: changes in the incidence of poverty and the level of income inequality

Region and area	Poverty as % of population 1962-3	1968-9	1975-6	1981	Gini coefficients* 1962-3	1968-9	1975-6	1981
North	65	36	35	23	0·359	0·370	0·422	0·456
Urban	56	19	31	23	0·460	0·440	0·453	0·462
Rural	66	37	36	23	0·308	0·345	0·368	0·422
North-east	74	65	46	36	0·344	0·379	0·405	0·438
Urban	44	24	38	36	0·422	0·450	0·457	0·456
Rural	77	67	48	36	0·264	0·347	0·343	0·395
Central	40	16	16	16	0·391	0·401	0·399	0·430
Urban	40	14	20	24	0·384	0·399	0·425	0·445
Rural	40	16	15	14	0·375	0·392	0·376	0·418
South	44	38	33	21	0·402	0·401	0·449	0·456
Urban	35	24	29	18	0·360	0·450	0·465	0·443
Rural	46	40	35	22	0·370	0·325	0·402	0·426
Bangkok	28	11	12	4	n.a.	0·412	0·398	0·405
Whole kingdom	57	39	33	24	0·441	0·429	0·451	0·473
Urban	38	16	22	16	0·405	0·429	0·435	0·447
Rural	61	43	37	27	0·361	0·381	0·395	0·437

*The larger the Gini coefficients, the more unequally income is distributed.

Source: National Statistical Office Household Expenditure Surveys.

Whatever the global, regional, and national trends, it is clear that while many Third World people have benefited from rural development programmes large numbers have not. Indeed, it is probable that some are worse-off. Recognizing these failures and investigating what lies behind them are more important than the debate over the number of poor people. Certainly, concern over the failure of programmes to reach or benefit the poorest groups has resulted in attention being focused on what might broadly be called 'alternative approaches'. These range from self-help community-based projects to proposals based on the experience of rural development in the socialist Third World countries. These tendencies have been reinforced by the shortage of development funds and the need to develop low-cost programmes.

Crop improvement and the farmer

The limitations of the Green Revolution were examined in the previous chapter. Since the mid-1960s HYVs have been developed that produce exceptional yields under ideal conditions. The success and spread of these varieties was governed initially by the degree to which farm conditions mirrored those prevailing on experimental stations. Subsequently the HYV have been modified to make them more suited to farm conditions. The limitation of this approach was revealed in case study J. Many critics of the 'technocratic approach' have argued that more progress would have been made if work had started from the actual situation prevailing on Third World farms. As a result there has been a growing appreciation of the appropriateness of indigenous methods of cultivation (chapter 2), selection of varieties (case study J), and knowledge of the local ecology (case study K). It is also the case that 'farmer-centred' approaches are very much cheaper than, for example, the Indian IADP described in chapter 6. In the prevailing international economic climate this has become an important consideration in rural development strategies.

The Green Revolution type of approach is also highly energy-intensive. Some have drawn attention to the energy efficiency of much traditional agriculture (figure 7.1). This has been a further line of argument for developing traditional systems of cultivation rather than replacing them. In addition the high-cost technocratic programmes impose change on the rural community from outside. Little account is taken of the community's views of problems and solutions. These may well be very different from those of the development planners (case study L).

Some writers consider that attempts to impose change from outside on the rural community can result in unsuitable, superficial, and

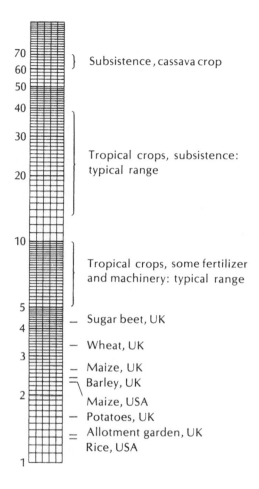

Figure 7.1 Energy ratios for Developed and Third World agriculture. The relationship between the energy input in agriculture and the output is sometimes known as the Energy Ratio. It is expressed as Energy output/Energy input (measured in calories). For Third World agriculture the ratio can be high as 70. Typically tropical subsistence crops range between 15 and 40. *Source*: Glaeser and Philips-Howard (1987: 183)

unsustainable development. For them successful development must originate within the community. This has given rise to the advocacy of 'bottom-up' approaches rather than 'top-down'. From this a major debate over 'development from above or below' has emerged.

Case study L

Farmers' and scientists' views of soil erosion

The differences between farmers' and technical experts' perceptions of problems can be enormous. In 1981, 750 shifting cultivators were interviewed in fifteen chiefdoms in Sierra Leone. The chiefdoms were then ranked according to the severity of the erosion hazard, based both on farmers' perception (table L1, column 1) and on environmental and socioeconomic measures suggested by previous scientific studies (table L1, column 2). The lack of correlation between the 'indigenous' and 'Western scientific' rankings is striking. Such findings cast serious doubt on the effectiveness of official soil conservation measures if they are targeted at 'erosion black spots' identified by the 'Western scientific' criteria.

(*Source*: Millington 1987: 16–19)

Table L1 Indigenous and Western scientific rankings of soil erosion in Sierra Leone

	Ranking of erosion risk			
		Western		
	Indigenous	scientific	Population	
Chiefdom	knowledge	knowledge	density*	Relief†
Lugbu	1	14	1	1
Tane	2	8	1	1
Toli	3	3	5	5
Njaluahua	4	15	1	1
Kunike Barine	5	1	4	5
Gallinas Perri	6	7	5	5
Lei	7	2	4	4
Sufrako Limba	8	5	3	3
Wara Wara Yagala	9	6	2	2
Ribbi	10	4	3	3
Small Bo	11	12	2	2
Dusse	12	9	4	4
Kari	13	10	2	2
Kpandakemo	14	11	5	4
Pejewa	15	13	3	3

*Population density class (people per square kilometre): *1* under twenty,, *2* twenty to twenty-five, *3* twenty-five to thirty-five, *4* thirty-five to sixty, *5* over sixty.
†Relief class: *1* under fifty metres, *2* fifty-one to 330 metres, *3* over 330 metres.

Source: Millington (1988).

In practice many programmes which claim to 'involve the community' in the development planning process do so only to a very limited extent. In other words the community is expected to choose from a list of options determined from above. Real 'grass roots' programmes where farmers, specialists, and planners work together (as in case study K) are all too rare.

Who is the farmer?

The tendency for rural development programmes to by-pass the poorer families and the subsistence sector has been widely reported (see chapters 5 and 6). Less attention has focused on who within the household should be targeted. The view that every household has a male head who is also the 'farm manager' is so deeply ingrained in conventional development planning that it is seldom actually stated. This is far from an accurate picture, particularly in much of sub-Saharan Africa and the Caribbean. In the former region, it has been estimated, as many as one third of rural households are headed by women. Many have no permanently resident adult male. Men are likely to be engaged in non-agricultural activities which may involve long-distance migration. In addition, even where there is an identifiable permanent male 'head of household', cultivation, particularly of subsistence crops, may well be a female activity. For much of Africa, women probably contribute 60 to 80 per cent of the labour and management that go into subsistence crop production.

In many instances the men are engaged in cash crop production and are the 'agricultural managers' targeted by rural development programmes. Traditionally men have engaged in non-agricultural activities such as crafts, hunting, and trading. These were displaced as incorporation into the market economy took place (chapter 3). In many areas the market economy comprised the production of export crops, and is described by Taylor (cited by Spring 1979: 332).

This one they call 'farmer'; send in teachers to teach him to farm (while I'm out growing the food); lend him the money for tractors and tillers (while I'm out growing the food); promise him fortunes if he'd only raise cotton (while I'm out growing the food); buy our land from him to add to your ranches (while I'm out growing the food).

For some writers the neglect of the female-dominated subsistence sector has played a major part in the emergence of the African food crisis. The policy implications of women's roles in the food system are summarized in figure 7.2.

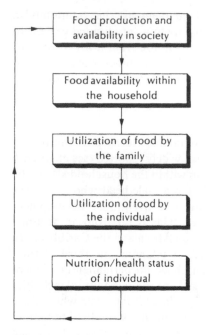

Low productivity, lack of access to credit, neglect by extension services, consumer orientated prices, lack of research into traditional crops,etc.

Limited access to cash (income generating schemes?)

Appropriate technology for storage and processing, improved water supply, conservation schemes (for wood) and alternative fuels?

Any time-saving innovations potentially increase women's time for childcare and leisure.

Any energy-saving mechanization potentially reduces requirements.

Figure 7.2 Policy issues arising from women's roles in the food system. *Source:* Trenchard (1987: 155)

Rural development in the 'socialist' Third World countries

A large and extremely varied group of Third World countries may be considered to exibit some form of 'socialism'. For a fuller discussion and some interesting case studies see Forbes and Thrift (1987) or Drakakis-Smith (1987). Although there is no space here to discuss the very different economies and societies of these countries, it is clear that they have had some notable successes in rural development, as well as, it must be said, some serious failures. The 'socialist' states of sub-Saharan Africa have been heavily criticized for the shortcomings of many of their rural development programmes. This is particularly the case with the Ethiopian resettlement and 'villagization' programmes. However, the failure of rural development programmes is by no means a monopoly of the socialist states. One of the most disastrous rural situations in the continent is found in pro-Western Zaire.

A particular focus of attention has been China, where a wide range of approaches to rural development have been tried (table 7.3). Whatever the

Table 7.3 Alternative approaches to promoting rural development in China

	Education	Land reform	Collectivization	Structural reorganization	Rural equity	Modernization
Perceived causes of rural underdevelopment						
	Illiteracy; shortage of production resources	Land concentration	Re-emergence and perpetuation of exploitative relations	Underutilization of resources	Unimproved productivity in agricultural production; lack of concentrated effort in grain production; insufficient public accumulation of funds for investment	Lack of incentives to increase production; rigid central planning; neglect of peasant cultivators' needs
Strategies						
	Mass education; basic literacy; agricultural technology; extension of credit; co-operatives	Redistribution of land; establishment of the principles of peasant ownership of land	Structural change through collectivization; collective production system	Mobilization of resources through organizational intervention; overcoming bottlenecks via labour mobilization; industrialization through agricultural development; moral/ideological incentives	Rigid central planning in agricultural production; greater moral/ideological incentives; grain production as the key link; concentrated effort in developing public/collective production sector; restriction on private and sideline production; overcoming capital bottlenecks by expansion of basic accounting unit (production unit); greater equity in distribution of work points	'Responsibility system'; accountability in economic management; emphasis on economic efficiency; agriculture as priority, then light industries, then heavy industries; local autonomy with respect to production and distribution of collective income; diversification of agriculture; low agricultural tax

Source: Ip and Wu (1983: 301–32).

failures and shortcomings of Chinese programmes, it is difficult to dismiss the very real achievements. The description of rural China in the 1930s (quoted in chapter 3) reflected a situation when, with a population of less than 500 million, famine struck somewhere in the country almost every year. Now, in the 1980s, with double the population, food shortages have largely been eliminated.

The socialist states have all experienced major problems of establishment and survival in a predominantly capitalist and often actively hostile world. Many already poor countries have been seriously weakened by protracted struggles. American interference in Nicaragua is merely the most recent example of a process seen at its most extreme in the Vietnam war. Scarce resources have in consequence had to be diverted into defence (in Vietnam this accounted for over 40 per cent of GDP in 1985), thus limiting the funds available for social and economic development.

Of necessity socialist Third World countries have attempted to 'disengage' from the world capitalist economy, while capitalist governments and businesses have put enormous pressure on them to re-engage. In general, they have been forced to rely heavily on limited domestic resources to promote development. This is most clearly illustrated in the case of Burma, where a policy of virtual isolation has been followed. However, eventually (in almost all cases) some form of accommodation with international capitalism has taken place, most recently and spectacularly in China, where 're-engagement' with international capitalism has been accompanied by radical changes in development policies. The need, in Premier Deng's words, to 're-establish the good names of prices and profits' has been followed in the rural areas by the abolition of the commune system. Land is now being cultivated as 'family farms' on fifteen-year leases.

The recent changes in China illustrate a major problem of Third World socialist development. Those countries remain, like the Third World as a whole, predominantly rural, with a weak urban industrial sector. While some have followed an actively pro-rural strategy, the attractions of urban industrial development have generally resulted in such policies being short-lived. In addition, the vestiges of rural communal organization in many, particularly African, countries have initially formed the basis for socialist structures. These have frequently proved a far from sound foundation. Indeed, it has been in the rural sector that private production for the market is most likely to have continued or first re-emerges in any retreat from socialist production.

In terms of measuring the 'success' of socialist rural development it

Case study M

The mechanization of Cuban sugar production

Prior to the 1959 revolution the majority of the Cuban rural population were employed as agricultural workers, but only 30 per cent were classified as farmers. The main source of employment was highly seasonal work on the sugar estates. Between 1959 and 1963 a series of measures resulted in major changes in the rural employment structure:

1 Between 1959 and 1963 agrarian reforms abolished rents in kind and cash for 100,000 small farms, confiscated large estates, and set up state farms.
2 Agricultural production was increased and diversified, particularly by combining underutilized land with underemployed labour.
3 State farms conferred the right of permanent employment on the previously seasonally employed.

As a result, by 1966 the number of agricultural workers had fallen by 20 per cent. This, combined with the diversification of agriculture, resulted in a serious shortage of seasonal cane-cutting labour, a sharp fall in sugar production (table M1) and a fall in export earnings.

Table M1 Cane-cutters employed in peak periods of the sugar harvest and percentage of cane harvested manually

Year	No. of cane-cutters ('000)	% of harvest cut manually	Total cane harvest (billion tons)
1970	350	99	82
1971	274	97	52
1972	210	93	44
1973	229	89	48
1974	200	82	50
1975	175	75	52
1976	153	68	54
1977	139	64	60
1978	153	62	70
1979	126	58	73

Source: Pollitt (1982).

Case study M *(continued)*

Temporary use was made of volunteer labour, the army, and students. Apart from the low productivity of such labour, it was recognized by the state that physical labour as extreme as cane-cutting should be reduced as far as possible. As a result a programme of mechanization was undertaken. However, serious problems were encountered in the development of combine harvesters. It was not until the mid-1970s that there was a breakthrough in mechanization and a steady rise in sugar production.

should be noted that many countries with a poor record of economic growth, as measured by *per capita* GDP, have performed much better on social indicators, such as mortality rates or life expectancy. This is certainly the case in Tanzania and, to an even greater extent, in Burma, and reflects, to a degree, different priorities.

Comparisons between socialist and non-socialist states (table 7.4) have resulted in much re-evaluation of capitalist development policies. There is little doubt that the development programmes of the socialist states have influenced development planning elsewhere in the Third World. However, the use of the socialist states almost as 'development laboratories' overlooks the fundamentally different nature of these regimes and the different approach to development that it implies (case study M). The

Table 7.4 Income distribution: Cuba, Brazil, Peru

| Country | % national income | |
	Lowest 40%	Upper 5%
Cuba		
1958	6·5	26·5
1978	24·8	11·0
Brazil		
1960	11·5	27·7
1980	9·9	34·2
Peru		
1961	10·0	26·0
1979	8·2	37·3

Source: Susman (1987).

underlying unity of the socialist countries comes from their varying degrees of anti-imperialism and policies of removing capitalist structures. Their rural development programmes cannot be evaluated outside this context. Is it realistic to contemplate transplanting, for example, elements of past Chinese rural development while keeping capitalist structures intact?

Key ideas

1 The degree to which rural development programmes have been successful is a matter of considerable debate. All too often evaluation is undertaken by those with a vested interest in declaring programmes a 'success'.
2 Changes in the depth and incidence of Third World rural poverty are far from clear. However, it is generally agreed that large numbers of the poorest people have received little benefit from development programmes.
3 Although most agricultural programmes are aimed at male farmers, in many cases food crops are cultivated by women.
4 Concern over the failure of programmes to reach the poor, coupled with a shortage of development funds, has resulted in questioning of 'orthodox policies'.
5 Increased attention has focused on 'farmer-centred approaches', the development of rain-fed cultivation, low-cost self-help schemes, and the development policies of the socialist Third World countries.

Review questions, further reading, and references

Chapter 1 The rural sector

1 Using table 1.1, compare a description of the household based on any *one* year with the picture that emerges from a study of all five years.
2 Using the latest World Bank *World Development Report*, describe the changing pattern of Third World grain imports.
3 Why are food production figures for Third World countries not necessarily a good indication of domestic food availability?

Bradley, P. N. (1985) 'Food production and distribution – and hunger', in R. J. Johnson and P. J. Taylor (eds), *A World in Crisis?*, Oxford: Blackwell.
Chambers, R., Longhurst, R., and Pacey, A. (eds) (1981) *Seasonal Dimension to Rural Poverty*, London: Pinter.
George, S., and Paige, N. (1982) *Food for Beginners*, London: Writers' and Readers' Publishing Co-operative.
Grigg, D. (1985) *The World Food Problem, 1950–1980*, Oxford: Blackwell.
World Commission on Environment and Development (1987) *Food 2000: Global Policies for Sustainable Agriculture*, London: Zed.

Chapter 2 Traditional rural production systems

1

	Wet year	Medium year	Dry year	Mean
Probability of conditions	0·33	0·33	0·33	–
Crop 1	20	15	10	15
Crop 2	150	35	0	47
Crop 3	15	15	60	35

The above table gives the production (in numbers of sacks) of three crops under different rainfall conditions. If they were basic food crops, which would a subsistence farmer cultivate? Would the choice be different if they were non-food cash crops?

2 Third World farmers have frequently been described as: not acting rationally; failing to respond to market forces; innately conservative; and resistant to change. How justified are these views?

3 How does population growth stimulate change in rural societies?

4 In case study B the agricultural system rests on extensive cultivation; long-term population growth was accommodated by clearing more land. Consider the likely results of the exhaustion of suitable uncleared land and continued population growth.

Klee, G. A. (ed.) (1980) *World Systems of Traditional Resource Management*, London: Edward Arnold.

Ruthenberg, H. (1971) *Farming Systems in the Tropics*, London: Oxford University Press.

Chapter 3 Rapid rural change

1 Making use of table 3.4, outline the problems likely to be encountered in studying long-term changes in Third World rural communities.

2 Making use of figure 3.1, describe the pattern of cash crop 'transplanting'. Why do you think this process took place?

3 Summarize the changing pattern of risks that farmers face when they become dependent on the production of crops for sale.

4 Describe how case study C fits into the pattern of world agriculture summarized in figure 1.7.

Dinham, B., and Hines, C., (1983) *Agribusiness in Africa*, London: Earth Resources.

Harrison, P. (1984) *Inside the Third World*, Harmondsworth: Penguin Books.

Timberlake, L. (1988) *Africa in Crisis*, London: Earthscan.

Chapter 4 Rural poverty

1 Regroup the 'causes' of rural poverty listed in table 4.2 into: underlying causes; immediate causes; and consequences.

2 Relate the views of Mexican rural development presented in case study E: first, to the various explanations of rural poverty outlined in chapter 4; and, second, to the types of rural development summarized in table 4.4.

3 With reference to the Sahel drought, produce brief summary explanations within the following categories: environmental; overpopulation; the operation of the national and international economies; and the interaction of the environment with the socio-economic structures.

Chambers, R. (1983) *Rural Development: Putting the Last First*, Harlow: Longman.
Redclift, M. (1984) *Development and the Environmental Crisis: Red or Green Alternatives*, London: Methuen.
Timberlake, L. (1988) *Africa in Crisis*, London: Earthscan.

Chapter 5   Redistribution of resources

1   It is often asserted that fragmentation of farm holdings is inefficient. Consider the implications of a consolidation programme for the agricultural system described in case study B. How would the introduction of irrigation change the situation?
2   Why might governments prefer programmes of land colonization rather than land reform? Why might poor families receive little long-term benefit from such colonization?
3   Under what conditions might successful land reform programmes take place? Under what circumstances might land reform be in conflict with policies aimed at raising agricultural productivity?
4   With reference to table 5.3 suggest why small farmers may not benefit from rural credit programmes.
5   What measures could be taken to reduce the effect on farmers of seasonal and annual price variations in crop prices?

Canlas, M., Miranda, M., and Putzel, J. (1988) *Land, Poverty and Politics in the Philippines*, London: Catholic Institute for International Relations.
Lehmann, D. (ed.) (1974) *Agrarian Reform and Agrarian Reformism: Studies of Peru, Chile, China, and India*, London: Faber & Faber.

Chapter 6   Increasing productivity

1   The development of double-cropped rice in the Muda irrigation scheme (Kedah, Malaysia) was followed by mechanization. This was considered necessary because of the need for rapid harvesting and land preparation if full advantage was to be taken of the irrigation. The most striking development was the introduction of combine harvesters, as the following account reveals.

By 1980 these huge machines, mostly owned by syndicates of Chinese businessmen, were harvesting 80–90 per cent of the paddy land in the Muda region. To appreciate the kind of technological leap involved here, one must realize that in the space of nearly three years the use of sickles and hand-threshing tubs, which have certainly not changed in over a century, was replaced by enormous machines especially tracked to travel over the clayey soils of Muda, which could cut and thresh five acres of paddy in less than an hour.
The corresponding loss of wage work was equally massive. Before the machines were introduced there were basically three paddy field operations

which represented significant opportunities for men and women to earn wages: transplanting, cutting, and threshing. The first two were largely women's work and the last men's work. For smallholders and landless labourers, who represented well over half of the village households, the cutting and threshing of paddy was the largest income earning opportunity of the season. At a single stroke most of this employment was eliminated. Two other opportunities for earning wages or gathering paddy were also largely eliminated. Prior to double-cropping, young men might earn as much as $M 100 [Malaysian dollars] a season by carrying sacks of paddy from the field to the larger bunds or to the roadside where they could be then transported to the rice mill or the owner's house. The combine harvesters eliminated much of this work by transporting the threshed paddy to the roadside where it is bagged. Secondly, the combine harvester has eliminated gleaning. Under traditional harvesting techniques, large piles of rice straw, containing many grains of paddy that had not been beaten off the stalks, would remain in the field. Women from the poorest families in the village would descend on the paddy fields after the harvest and glean what they could. A week or so of gleaning might yield three of four large sacks of rice, which constituted an important food supplement for poorer villagers. The combine-harvesters grind up and destroy the paddy stalks, which are then distributed across the surface of the field, making gleaning impossible.

(*Source:* Scott 1984: 164)

Consider the implications of this situation for (*a*) large land owners, (*b*) landless labourers, (*c*) national food supply, and (*d*) rural poverty.

2   The Green Revolution has resulted in a number of Asian countries, for example Malaysia, Indonesia, and India, becoming virtually self-sufficient in basic foodstuffs. What are the implications of this for other Third World countries heavily dependent on grain exports?

3   Application of artificial fertilizers

	1974 kg/ha	1984 kg/ha
Asia	36	74
South America	24	26
Africa	13	13
All Third World	23·2	64·4
All Developed World	98·6	116·4

Source: World Bank (1987).

Using the above table and the material in chapter 4, comment on the relationship between the changing pattern of fertilizer use and the spread of HYVs.

4   Making use of figures 6.4 and 6.5, consider the extent to which a high level of irrigation reflects water shortages.

Barrow, C. (1987) *Water Resources and Agricultural Development in the Tropics*, Harlow: Longman.

Chapter 7 Assessments, achievements, and alternative approaches

1 Compare the trends in rural poverty shown in table 7.1 with those of *per capita* food production illustrated in figure 1.6.

2 Why, in recent years, have 'orthodox' approaches to rural development been increasingly questioned?

3 Using table L1, examine the relationships between the two rankings of erosion risk, population density and relief.

4 The following is a check list of questions designed to assist in the assessment of possible lines of rural development:

Productive activities
1 What do village people produce?
   (a) farm crops;
   (b) animal products;
   (c) produce from hunting and gathering: e.g. meat, honey, firewood, timber, medicinal ingredients, palm wine;
   (d) manufacturing and processing: milling activities, food processing, brewing and distillation, textiles, blacksmithing, basketry, brick-making, pottery etc. (including part-time activities for household use, as well as activities by specialist craft workers);
   (e) services: local medical specialities, midwifery, dispute settlement, divination, ritual expertise, music, carving etc.
2 Who produces what?
   Distinguish between major activities for men and women, and between activities which might reasonably be undertaken by each household from those requiring the service of specialists.
3 Access to productive resources?
   What is needed to start up each of the productive activities listed? How do individuals qualify for rights to land or hunting and gathering opportunities?
4 What happens to the product?
   (a) items produced on a household subsistence basis;
   (b) products and services entering into local exchange;
   (c) exports to the national or international economy.

Local skills
The aim of this section should be to present an inventory of local skills, as a framework for identifying both untapped 'self-help' potential and the crucial skill shortages.
Facilities, equipment and raw materials
   1. Transport.
   2. Tools, machines and household equipment.
   3. Water supply.
   4. Raw materials.
Social and political organization
   1. Organization of labour.
   2. Political organization.

3. Local perceptions of development priorities.

(*Source:* Richards, 1985)

(*a*) Why do you think this list contains topics that appear to be outside those normally considered by agricultural development agencies?

(*b*) How does this approach compare with those outlined in chapter 6?

5 What are the consequences of assuming that all farmers are male?

6 What lessons for the rest of the Third World may be learned from the experiences of the socialist countries?

Dankel, I., and Davidson, J. (1988) *Women and the Environment in the Third World*, London: Earthscan.

Drakakis-Smith, D. (ed.) (1987) 'This changing world: socialist development in the Third World', *Geography* 72, 4: 333–63.

Forbes, D., and Thrift, N. (eds) (1987) *The Socialist Third World*, Oxford: Blackwell.

Richards, P. (1985) *Indigenous Agricultural Revolution*, London: Hutchinson.

General references

Baker, R. (1982) *Land Degradation in Kenya: Economic or Social Crisis?*, Development Studies Discussion Paper No. 82, Norwich: University of East Anglia.

Barr, T. N. (1981) 'The world food situation and global grain prospects', *Science* 214: 1087–95.

Blaikie, P. (1985) *The Political Economy of Soil Erosion in Developing Countries*, Harlow: Longman.

Bray, F. (1986) *The Rice Economies*, Oxford: Blackwell.

Byres, T. J., and Crow, Ben, with Mae Wan Ho (1983) *The Green Revolution in India*, Open University course US 204, Third World Studies, Milton Keynes: Open University.

Dickenson, J. P., Clarke, C. G., Gould, W. T. S., Prothero, R. M., Siddle, D. J., Smith, C. T., Thomas-Hope, E. M., and Hodgkiss, A. G. (1983) *A Geography of the Third World*, London: Methuen.

Dixon, C. J. (1978) 'Settlement and environment in north-east Thailand', *Journal of Tropical Geography* 46: 1–10.

Douglas, M. (1983) 'The Korean Saemaul Undong: accelerated rural development in an open society', in D. A. M. Lea and D. P. Chaudhri (eds), *Rural Development and the State*, London: Methuen.

Fitzgerald, M. (1978) *Drought, Famine and Revolution in Ethiopia*, Occasional Paper No. 1, London: School of Oriental and African Studies, University of London.

Food and Agriculture Organization (1967) *Production Year Book* 21, Rome: FAO.

— (1977) *Production Year Book* 31, Geneva: FAO.

— (1987) *Production Year Book* 41, Rome: FAO.

Frankel, F. R. (1971) *India's Green Revolution: Economic Growth and Political Costs*, Princeton, N. J.: Princeton University Press.

Glaeser, B., and Philips-Howard, K. D. (1987), 'Low energy farming systems in

Nigeria', in B. Glover (ed.), *The Green Revolution Revisited*, London: Allen & Unwin.

Griffen, K. (1974) *The Political Economy of Agrarian Change*, London: Macmillan.

Grigg, D. (1984) *An Introduction to Agricultural Geography*, London: Hutchinson.

Guthrie, C. (1986) 'The African environment', in A. Hansen and D. E. McMillan (eds), *Food in sub-Saharan Africa*, Boulder, Col.: Lynee Rienner.

Hardjono, J. (1983) 'Rural development in Indonesia: the "top down" approach', in D. A. M. Lea and D. P. Chaudhri (eds), *Rural Development and the State*, London: Methuen.

Harrison, P. (1981) 'The inequalities that curb potential', *Ceres* 81: 22-6.

Heibsch, C., and O'Hair, S. K. (1986) 'Major domestic food crops', in A. Hansen and D. E. McMillan (eds), *Food in sub-Saharan Africa*, Boulder, Col.: Lynne Reinner.

Hopkins, M. J. D. (1980) 'A global forecast of absolute poverty and employment', *International Labour Review* 119: 565-79.

Hunt, D. (1984) *The Impending Crisis in Kenya; the Case for Land Reform*, London: Gower.

International Labour Office (1974) *Sharing in Development*, Geneva: ILO.

Ip, D. F., and Wu, Chang-Tong (1983) 'From subsistence to growth: Chinese strategies of rural transformation', in D. A. M. Lea and D. P. Chaudhri (eds), *Rural Development and the State*, London: Methuen.

Leite, L. L., and Furley, P. A. (1985), 'Land development in the Brazilian Amazon', in J. Hemming (ed.), *Change in the Amazon Basin* II, Manchester: Manchester University Press.

Millington, A. (1988) 'Local farmer perceptions of soil hazards and indigenous soil cultivation strategies in Sierra Leone, West Africa', in I. P. Sentis (ed.) *Soil Conservation and Productivity*, Proceedings of the International Conference on the Conservation of Soils, Maracay, Venezuela.

Open University (1982) *Atlas of the Third World*, Milton Keynes: Open University Press.

Pollitt, B. H. (1982) 'The transition to socialist agriculture in Cuba', *Institute of Development Studies Bulletin* 13, 4: 12-22.

Raikes, P. (1981) 'Seasonality in the rural economy', in R. Chambers, R. Longhurst, and A. Pacey (eds), *The Seasonal Dimension in Rural Development*, London: Pinter.

Sarap, K. (1987) 'Transactions in rural credit markets in Western Orissa, India', *Journal of Peasant Studies* 15, 1: 83-107.

Scott, J. C. (1976) *The Moral Economy of the Peasant*, New Haven, Conn.: Yale University Press.

— (1984) 'History according to winners and losers', in A. Turton and S. Tanabe (eds), *History and Peasant Consciousness in Southeast Asia*, Osaka: National Museum of Ethnology.

Spring, A. (1979) 'Women farmers and food in Africa', in A. Hansen and D. E. McMillan (eds.), *Food in sub-Saharan Africa*, Boulder, Col.: Lynne Rienner.

Susman, P. (1987) 'Spatial equality and socialist transformation in Cuba', in D. Forbes and N. Thrift (eds.), *The Socialist Third World*, Oxford: Blackwell.

Tawney, R. H. (1932) *Land and Labour in China*, repr. Boston, Mass.: Beacon Press, 1966.
Taylor, D. C. (1981) *The Economics of Malaysian Paddy Production and Irrigation*, Bangkok: Agricultural Development Council.
Trenchard, E. (1987) 'Rural women's work in sub-Saharan Africa and the implications for nutrition', in J. Momsen and T. Townsend (eds), *Geography and Gender in the Third World*, London: Hutchinson.
United Nations (1987) *Environmental Data Report*, Oxford: Blackwell.
United Nations Conference on Trade and Development (1987) *Handbook of International Trade and Development Statistics*, Geneva: UNCTAD.
Webster, D. J. (1986) 'Food production and nutrition in southern Africa: historical perspective', *Modern African Studies* 23, 3: 447-63.
World Bank (1985) *Rural Development Sector Policy Paper*, Washington, D. C.
— (1987) *World Development Report*, Oxford: Oxford University Press.
— (1988) *World Development Report*, Oxford: Oxford University Press.

Index